Technology in school

Routledge Education Books

Advisory editor: John Eggleston
Professor of Education
University of Warwick

Technology in school
A handbook of practical approaches
and ideas

John Cave

Routledge & Kegan Paul
London, Boston and Henley

First published in 1986
by Routledge & Kegan Paul plc

14 Leicester Square, London WC2H 7PH, England

9 Park Street, Boston, Mass. 02108, USA

Broadway House, Newtown Road,
Henley on Thames, Oxon RG9 1EN, England

Set in Times, 10 on 11pt
by Columns of Reading
and printed in Great Britain
by T.J. Press (Padstow) Ltd
Padstow, Cornwall.

Library of Congress Cataloging in Publication Data

Cave, John, 1949–
Technology in school.
(Routledge education books)
1. Technology – Study and teaching (Primary)
2. Technology – Study and teaching (Secondary)
I. Title.   II. Series.
T65.3.C39   1985   372.3'5   85-14246

British Library CIP data also available

ISBN 0-7102-0528-7
ISBN 0-7102-0732-8 (pb)

Contents

Contents

Foreword

Technology is fast becoming a key subject in the curriculum of secondary and primary schools as teachers endeavour to prepare young people for life in a technological society. But all too often the technology that is taught is a distorted and incomplete version of industrial technology, inappropriate for the needs of young children, for the resources of the schools and the expertise of the teachers.

John Cave has devised a new and practical way of teaching the principles and processes of technology that is geared to the capabilities of children and their schools. Using inexpensive materials and equipment his approach enables young people not only to learn about technology effectively but, even more importantly, to adopt a technological perspective on work and leisure. Even though the approach is simple and direct it leads even young children to the very frontiers of modern technology.

John Cave calls his approach 'appropriate technology'. It has already enhanced the curriculum of many schools and inspired the learning of many young people. In this book the further development of his ideas will ensure that this work will be enhanced and reach an even wider audience. The result will be both to transform and establish more fully the place of technology in the lives of schools and their students.

John Eggleston

Preface

The material for this book has been collected together and developed in close liaison with Craft, Design and Technology (CDT) specialists during the last four years. It is presented here in the hope that other colleagues may find some, or even all of it, of interest and of value in producing teaching material of their own. It is very much a book for teachers of CDT, and it therefore makes reasonable assumptions about the reader's knowledge of materials, processes, etc. Since space is at a premium, the section on microelectronics necessarily makes assumptions too, but as many concessions as possible have been made to those less familiar with this area.

All additional notes, addresses of suppliers and suggested further reading can be found at the end of each section. The information is as comprehensive as possible and should enable the reader to get a firm foothold into the proposed teaching activities.

In preparing this book, I have been fortunate enough to enjoy the co-operation of many CDT colleagues who have tried out ideas in school and suggested improvements. It confirms my belief that the development of this kind of material should be a joint enterprise transcending and (hopefully) demolishing some of the older institutional barriers between teacher education and teaching in school.

I am grateful for the stimulus and support given by colleagues at Middlesex Polytechnic and the opportunities offered through DES and LEA in-service courses for the dissemination and trialling of ideas in schools. In particular I must thank those who have allowed me to illustrate their work: Keith Good of Ashford Boys' School, members of the Hampshire in-service course 85/9 (1984) for CDT who took the option on low-cost modelling;

Preface

members of Hampshire in-service course 85/39: Mr G. Gold-
thorpe, Mr P. Lewin; students of Middlesex Polytechnic: Geoff
Pragnell, Cathy Osbond, Rick Law, Duncan Taylor, Bill Dexter,
Michael Dench, Jeremy Burrowes, Barry Gillespie, Karen
Allaway, David Allibone, Howard Baxter, Finola Byrne, Gillian
Coxshall, Judith Friedner, Henry Hawes, Clive Hollamby,
Naomi Johns, David Parry, Fred Vintner.

In addition, I must also thank the following: the technician
staff in 'D' Block at Middlesex Polytechnic, especially Wally
Scott who has helped in making equipment; David Prest for his
introduction to corrugated plastic sheet and its potential; Peter
Chambers for his introduction to the Velos eyeletting system;
Richard Tufnell (and David Prest) for the benefit of discussion;
Margaret Emmerson, who has trialled much of the material of
section 1 in school.

Finally, I am especially grateful to Professor John Eggleston
whose idea it was to assemble this material between two covers,
and who has provided every encouragement for the work through
conferences at Keele University and in the pages of *Studies in
Design Education, Craft and Technology*. Grateful acknowledg-
ment is also made for permission to reprint parts of the original
articles appearing in that journal.

Introduction

As its title suggests, this is intended to be a practical resource book for CDT specialists; it makes certain assumptions about the nature of CDT, but there is no attempt to theorise at great length about this.[1] Rather, the book aims to show through example – in a very practical and down-to-earth way – how elements of technology can be incorporated into and developed within existing programmes covering lower-school through to sixth-form work.

Section 1 opens with a range of proposals for foundation work in CDT. Although *modular courses in technology* have now gained a firm foothold in many schools, there is so far very little provision for technology in the lower years. This state of affairs may well improve rapidly during the next few years, but any developments will have to take place in a climate of very tight cash constraints and limited resources.

A series of short articles in the journal *Studies in Design Education, Craft and Technology* used the expression *appropriate technology* (in the context of school technology) to describe a number of ideas for getting technology into lower-school work cheaply and using existing resources as far as possible.[2] Extensive trialling in schools has since confirmed that several important elements of technology – e.g., pneumatic and hydraulic control principles – can be taught in an effective, stimulating and enjoyable way through designing and making.

The original articles from SDECT are reprinted here in a revised and enlarged form, and much new material has been added. In particular, some attention has been given to the problem of introducing simple electrical/electronics work into lower-school CDT. This is a sadly neglected topic, and especially worrying at a time when children themselves are becoming increasingly aware from a very early age of the significance of

1

electronics through computing, games and the use of all kinds of consumer goods.

One important area not covered by a specific technology module in the Schools Council scheme (but implicit in several) is *production technology* – the subject of *Section 2*. This makes no attempt, of course, at a comprehensive coverage of production methods, but instead looks at several important commercial production techniques often thought to be out of reach of CDT specialists because of the cost of equipment, its complexity – or both. There are in fact many fascinating commercial techniques for forming materials which can be effectively presented to children through the use of simple improvised equipment. And not only can techniques be demonstrated, the same equipment can be put to practical use to form materials in ways not otherwise possible in school.

A good example of a relatively unusual production method, conducive to the appropriate technology treatment, is hydraulic forming. The simple equipment for doing this can be constructed in a very short time, and is capable of remarkable feats of metal deformation. Once the former, shown in *Section 2*, is available, pupils of all ages can design and make suitable dies without resort to any machine and manufacture sheet metal pressings in any quantity. Similarly, pupils can learn about chemical machining, cryogenic forming and many others through practical involvement with the techniques themselves.

Section 3 of the book is also concerned with manufacturing techniques, but specifically in the context of prototype model making. The nature of project work in CDT – especially coursework for examinations up to 'A' level – is rapidly changing, and if any single new trend can be picked out it is probably the shift away from more traditional 'one-off' designing and making to product design work having something of a commercial flavour. Increasingly common at 'A' level, for example, are those assignments involving the design of products capable in principle of mass-production – say, by injection moulding. These naturally make rather different demands and, among other things, require some knowledge of how to represent ideas in an appropriate three-dimensional model form.

Very exact product modelling – i.e. making the model look like the real thing – is familiar enough in the commercial world of industrial design, but has yet to become well established in CDT. The aim here is therefore to introduce a range of exact modelling techniques suitable for use in school where resources may be severely limited.

It will become quite clear that all the product models illustrated in *Section 3* are made to *function* in some important respect as well as to show the visual and other qualities of the intended product. For sixth-form students (and younger pupils) nothing seems to motivate quite like the prospect of a prototype model with a highly professional look – and which *works*.

It is not always possible, of course, to end up with a fully functioning model; it depends entirely on the nature of the project in hand. Those entailing complex mechanical elements, for example, present obvious difficulties, but others can be very suitable indeed. In particular, this book draws attention to the many possibilities for designing and modelling small battery-powered products such as torches, timers, measuring devices, alarm units, etc.

There is clearly an excellent opportunity here to arrange a suitable marriage between mainstream designing and making and electronics. The title given to *Section 4* of the book – *Microelectronics for Product Design* – really sums up its intention; it is not so much concerned with microelectronics *per se* as with the application of cheap microelectronic devices and circuitry to product design work. An incalculable number of commercial products now depend on microelectronics, and against this background many 'A' level students (and CDT specialists) are becoming increasingly frustrated when solutions to product design assignments call for electronic circuitry to provide particular functions such as timing or electronic switching.

If a marriage between product design work and electronics can be sustained, there is an obvious need for a minimal grasp of the potential of microelectronics here – plus some tangible help in providing useful and relevant circuit ideas which can be used either as the basis for assignments or for adaptation to particular needs as they arise.

Section 4 examines a number of circuit ideas which seem to have an obvious cash value in product design work. One only has to examine products such as video recorders, for example, to realise that a good deal of switching is now accomplished electronically to provide the latching – say – for push-button 'on'/'off' control. The first circuit discussed provides the means for simple electronic switch-latching – which also makes it suitable for alarm systems, games, and many more applications besides. It costs only a few pence to assemble, requires only three components (including the integrated circuit), but could probably sustain an 'A' level course for years in terms of the number of

3

projects that might be generated around it.

It might seem at first glance that this book is something of a miscellany with little obvious coherence between the four parts. This is true to the extent that the material in each section was produced at different times for specific reasons, and appears here in a collected form. Although it should be possible to make use of each section quite independently of any other, it will become clear to the reader that much of the material is inter-related and that several themes recur throughout and underpin the thinking behind all of it.

The most important theme of all, perhaps, is the conviction that technology can be taught as a component of CDT to children of all ages and different abilities through mainstream designing and making. Although it may be necessary to specialise for examination purposes, I can see no good reason for wanting to strip technology away from CDT in the lower school; indeed, if CDT is the shorthand for a unified activity, it would seem less than honest to do so.

I am also strongly convinced of the need for ideas which have an almost immediate cash value for those wishing to build technology into an existing or new course structure. This entails realism in suggesting any kind of project work, especially a clear recognition that resources are scarce and that CDT specialists everywhere are having to provide for increasing numbers of children with less and less time for the preparation of teaching materials. What is offered here is no panacea but a set of proposals based on a genuine appreciation of the problems – and healthy common sense.

Finally, technology with a capital 'T' makes inexorable progress, and there is a constant need to keep teaching material up to date. To translate even a fraction of recent developments into relevant teaching activities would be a limitless and unmanageable task; however, as the book attempts to demonstrate, it is possible to be highly selective and introduce even very young children to subject matter quite close to the frontiers of technology itself.

Notes

1 For an excellent recent analysis see: B.K. Down, 'Educational Aims in the Technological Society', *Studies in Design Education, Craft and Technology*, vol. 16, no. 2, 1984, pp. 68-74.
2 *Studies in Design Education, Craft and Technology*, vol. 13, no. 1, 1980,

vol. 13, no. 2, 1981, vol. 14, no. 2, 1982. See also: J.F. Cave, 'CDT in the Middle School', *Practical Education*, vol. 80, no. 3, 1982.

Section 1

Getting technology into CDT

The general case for a fresh appraisal of technology in lower school work has been set out very briefly. If the argument is accepted, one of the problems – if not the central problem – becomes one of creating specific briefs, assignments or projects for particular groups at particular times. What follows here is not intended to be representative of an assortment of ready-made briefs or 'jobs', but to illustrate a few ways in which important aspects of technology can be introduced to younger children through designing and making. Taught this way, there is now plenty of evidence to show that quite sophisticated ideas can be introduced at an early age – and make an indelible impression on young minds.

Control technology: pneumatics

Pneumatic control systems are basically of two types: *open* or *closed*. The first of these is without doubt the most useful commercially, but the requirements include a high-pressure air supply and expensive specialised components. Closed systems, on the other hand, although they have a comparatively limited range of applications, can be assembled very cheaply using little more than household throw-outs. A very simple system can be modelled using cheap PVC tubing[1] and the parts shown in fig. 1; when the bottle is squeezed, the balloon expands and is capable of a considerable lifting effort.

This type of arrangement is hardly original; there are plenty of examples of Victorian toys incorporating rubber bulbs and tubing to facilitate control over a distance. But this type of system has also been used for some time in the wider context of pneumatic

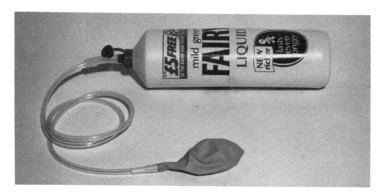

1 Balloon/bottle pneumatic control system

control. Among current examples, it is worth drawing attention to the pneumatically operated switch shown in fig. 2. When the bellows are depressed, the pressure increase causes deflection of a diaphragm to actuate a switch. Similar systems are common in light machinery – e.g., printing and duplicating machines – to effect light control movements from 'A' to 'B', replacing the complex mechanical linkages that might otherwise be needed. These systems are reliable, cheap, and they can make the designer's job a much easier one.

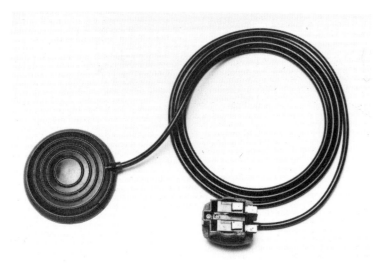

2 Pneumatically controlled switch

Using the balloon and bottle system, a useful control principle can be demonstrated very effectively. If a small pile of books is placed over a deflated balloon, they are easily toppled over when the system is pressurised. This experiment can also be turned into a quantitative examination of the system's potential; the surface area of the balloon presented to a load simply has to be multiplied by an assumed pressure in the system to give a figure for lifting capacity. Conversely, the pressure in the system can be calculated when the load and balloon area are both known.

It does in fact pay to look at the system in some detail before attempting to apply it in project work. Because the air is compressible there is a certain amount of 'spring' or resilience in the system, and it is unsuitable, therefore, where an immediate, crisp control response is required. The disposition of the balloon is important since the maximum area possible should be presented to the load or component being moved. If it is restricted in a narrow tube, for example to move a jack-in-the-box figure up and down, high pressures will be required to create much movement. And if the wall of this tube happens to be weak, it can easily be split apart if the area of contact with the expanding balloon is appreciable. Also unexpected, perhaps, is the problem of inflating two balloons from one bottle to produce two simultaneous control movements. Because of differing surface areas and the behaviour of the rubber itself, one balloon will always expand at the expense of the other.

Notwithstanding these problems, the balloon-bottle system can be incorporated into a wide range of projects necessitating some form of remote control. Figs 3-10 illustrate a number of variations on the theme of designing and modelling building construction plant. The problem here was to design and make a toy or model to represent a piece of earth-moving equipment with a single part moving under pneumatic control.

The brief for this particular project might be left very open for older pupils or sixth-form students (e.g. design of a toy for younger children), but it obviously requires to be more tightly structured for lower-school work. The examples shown are therefore based on a simple chassis and balloon-lift mechanism which would be similarly constructed by all and then built up through individual endeavours into any one of a wide variety of vehicle types. In other words, a proportion of the projection is tightly prescribed, but a significant amount of decision making is left to pupils in adding to the basic construction. The end result is a distinctive toy or model with a facility for action-at-a-distance control which invests such a project with an interest for most

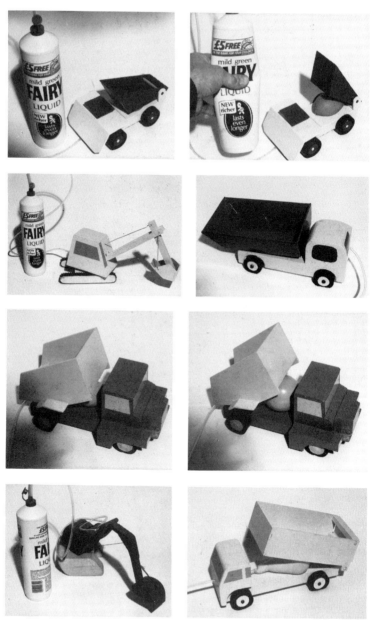

3/10 Earth-moving toys using the simple balloon/bottle pneumatic control system

children out of all proportion to the additional cost.

The simple bottle-balloon control system has been tried out in a large number of schools, and many other examples of applications come readily to mind: jack-in-the-box toys, games, animal animations, puppet animations, etc. The list is virtually endless; design briefs, specifying the use of pneumatic systems, can be framed to suit very different levels of thinking, and although the approach may be a purely qualitative one, the opportunity does exist for introducing simple quantitative considerations of how the system performs.

The materials and constructional methods used to produce the models illustrated are not themselves without interest. They could be put together using traditional materials and joining techniques, but here the main structural material is the 3mm thick chipboard originally developed for use in furniture and now extensively used for lining floors prior to carpeting.[2] This is by far the cheapest composite sheet material available, and its low cost stretches limited resources a long way. It is an excellent material to cut, pin and glue and is far easier for children to work with than hardboard – as well as being cheaper. And there is an incidental advantage in insisting on the use of 3mm chipboard as the principal material for a project such as the one described. The use of inherently weak sheet material itself creates problems that demand structural solutions, providing an opportuity for teaching about box sections, hollow-core construction, ribbing, etc.

Where a piece of work – or even the prototype for it – has to be assembled quickly, 3mm chipboard can be pinned and glued in combination with softwood sections; alternatively, use can be made of hot-melt glue. The wide availability of cheap hot-melt glue guns and adhesives now provides the possibility of almost instant three-dimensional modelling of ideas, and providing they are used appropriately and with due caution, they are an exceedingly useful resource.

Control technology: hydraulics

The pneumatic system described would seem to have much to recommend it as a means of adding an interesting dimension to some project work, but it does have severe limitations as a method of control (reflecting the limitations of similar closed systems in commercial use). This is not so much of an objection, however, in the case of hydraulic control systems based on the use of cheap disposable syringes interconnected with PVC tubing. The availability of these syringes and other inexpensive fittings has opened up a whole new range of possibilities for control work in CDT.[3] Although they are only single-acting, syringes can be incorporated in water-filled systems in a limitless variety of ways to represent important commercial applications. And precisely because these components are so cheap, it is now possible to think in terms of consumable hydraulic systems for project work.

There can be few more meaningful introductions to the concept of hydraulic control than the simple exercise shown in fig. 11. By taking a pair of interconnected syringes in hand and alternately pushing down one plunger against a 'load' imposed on the second, it quickly becomes clear to most children that not only can the system effect a control movement from 'A' to 'B' with little resistance, but the path of transmission is infinitely flexible, and can be extended simply by adding more tubing. As a system relying on a fluid, most children are surprised that it is so positive – any residual 'spring' being due to the elasticity of the PVC tubing. Similar problems caused by air entrapment can be examined by introducing a bubble of air and comparing the 'spongy' performance with the original positive action.

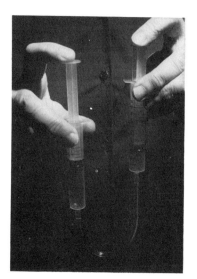 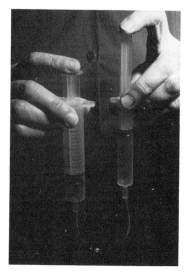

11 (above left) First experiment with a simple hydraulic control system

12 (above right) Obtaining a mechanical advantage with a simple hydraulic system

If the cross-sectional areas of the two syringes are dissimilar – say, in the ratio of 2:1 – the idea of obtaining a mechanical advantage by hydraulic means becomes *intuitively* obvious when the same experiment is tried (fig. 12). Using the same syringes, it can be shown how mechanical advantage is related to plunger cross-sectional area and how this in turn relates to length of stroke. A simple piece of apparatus for doing this consists of a block of wood drilled to accommodate the pair of syringes standing vertically side by side with the plunger tops adapted to receive weights. A more satisfactory version involves the addition of weight-hung pivoted arms, one acting on each plunger top. When these are appropriately loaded, any deviation from the expected theoretical result will arise from friction in what might be no more than a piece of makeshift apparatus, and the tightly fitting seals in the syringes.

Many more complex systems can be assembled with just one or two additional components. Fig. 13 illustrates a *master* cylinder and two *slaves* providing for simultaneous dual control; more slaves could be added – for example, to represent a car braking system.

A simple, but most effective, powered system can be made up by adding a standard electric windscreen washer pump. Typically, these are capable of delivering up to 40psi, and when water at

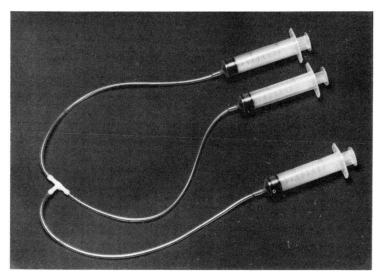

13 Two slaves operated from one master syringe

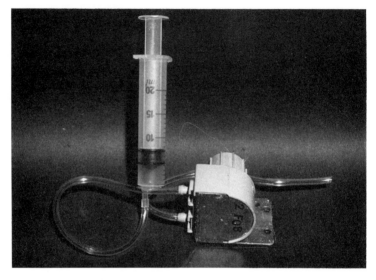

14 A simple powered hydraulic system using a windscreen washer pump

this pressure is supplied to a larger syringe an impressive performance is possible – in raising heavy loads, for example, or operating a further mechanism (fig. 14).

Again, the possibilities for building these hydraulic systems into models, toys and games seem endless. Although syringes are not double-acting, atmospheric pressure is normally sufficient to effect the return stroke when the plunger of the controlling syringe is pulled back rather than pushed down; additionally, a return spring can be built in at the slave end to assist return (e.g., a rubber band).

In designing and making models or toys, the theme of earth-moving equipment, discussed above, is particularly apposite because of the use of hydraulics in actual plant. Using the common chassis principle, for example, younger children can design and produce dumper trucks, bulldozers, forklift vehicles and many other variations besides.

Unlike the simple pneumatics, work with these hydraulic systems can be developed and put to more advanced uses higher up in the school. Figs 15-22, for example, illustrate a range of solutions to the interesting problem of handling small objects from a distance – i.e., the design and construction of a remotely controlled 'robot' arm. (There is in fact a well-established requirement for devices capable of manipulating toxic materials in hermetically sealed enclosures providing greater insulation to the operator than a glove box, but avoiding the high cost of an electric servo system.)

The problem set here was to design and make a device capable of controlled manipulation and movement of a specified object – the operator standing at some distance. Three hydraulic systems, comprising three pairs of syringes and a length of PVC tubing, were supplied for each device on the assumption that at least three principal control movements would be necessary to solve the problem. All the solutions shown originated from the same in-service course for CDT specialists, and they are especially noteworthy in illustrating the diversity of solutions in general, and a wide variety of different answers to particular problems. This kind of project can also be set as an assignment in school at any one of several different levels as an introduction to robotics, hydraulic control, elementary kinematics – and even structures. It is unusually rich in the experience(s) it offers, and is virtually guaranteed to create and sustain a good deal of interest.

Syringe hydraulic systems are invaluable too for other proto-type work. There are many situations – for example, in 'A' level project work – where any tentative thoughts about incorporating hydraulics in a design solution might otherwise be ruled out on grounds of cost or availability of suitable parts. These systems are extremely durable and robust, and will accept surprisingly high fluid pressures – failing only when the plunger stem distorts under load. And it may be worth underlining that as a means of getting a control movement from 'A' to 'B', a pair of syringes can be far cheaper than even a simple cable link.

Warning: Since the possibility of abuse, however remote, might be thought to exist, it is suggested that any syringes used for hydraulics work should be subject to a strict system of accounting and that any syringe taken away should be permanently built into a product. Some syringes are supplied with needles, and these should be destroyed immediately.

Elementary circuitry

It would have been somewhat grandiose and misleading to title this sub-section *electronics* since only one topic is discussed here as a contribution to the development of electronics foundation work.

If CDT can make a distinctive contribution in this area, then

15

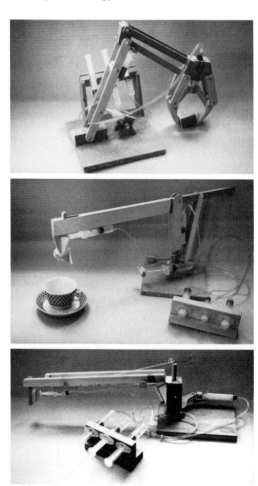

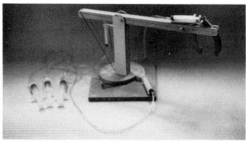

15/22 Remote control 'robotic' arms using hydraulic control systems based on syringes

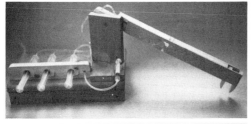

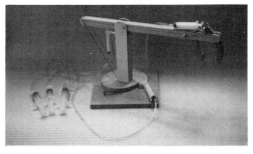

arguably it will be through the medium of project work that introduces basic circuit concepts and at the same time provides a rich and meaningful design and make experience. One way of doing this is to begin by looking at the problem of switching, and the application of a few switch types in very simple continuity circuits.

During introductory – or, for that matter, more advanced – circuit work, pupils are seldom given the opportunity to design and make switches; these are usually bought in for a particular purpose, and taken for granted by those who use them. But all kinds of switch can be designed and put together without much difficulty and, paradoxically, a switch type with many 'high technology' associations is the one that even very young children can design and make for themselves to incorporate into very simple continuity circuits.

Membrane panel switches are a laminate of thin flexible plastic sheets at least two of which are printed with a pattern of conductor tracks. These are typically separated one layer from another by the thickness of an insulating laminate having a corresponding series of windows or apertures stamped out of it; when pressure is applied at these points on the membrane surface, the top layer is deflected locally through the middle laminate window(s) to make electrical contacts (fig. 23).

The three layers of a very simple membrane panel switch are illustrated in fig. 24, and when assembled this would represent a SPST (single pole single throw) switch. It functions in the following way: Two parallel conductor tracks exist on the bottom

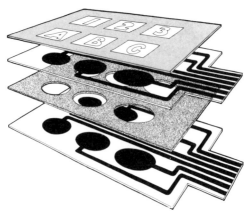

23 A typical multi-function membrane panel switch

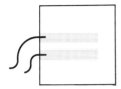

24a Membrane panel base *layer with two parallel conductor tracks*

24b Membrane panel centre layer with circular window cut out

24c Underside of top *membrane layer showing conductive area in the centre*

layer, and these terminate in flexible leads. This is overlaid with an insulating panel having a small window cut out of the centre. A third layer, with a conductive area in the centre of its underside, is placed on top. Normally, the conductive area on the top layer is separated from the tracks on the bottom by the thickness of the middle laminate, but when the top is depressed in the centre it deflects locally and the narrow gap between the conductors is bridged. When pressure is removed, the switch returns to its original condition.

This is a very simple version of a membrane panel switch; commercial varieties are typically designed and produced as multi-functional switches with a brightly printed top layer indicating pressure points (fig. 25). Membrane panel switches are already extensively employed in calculators, computer and cash register keyboards, vending and photocopy machines, etc. They have a number of advantages over conventional mechanical switches; they are cheap, reliable and give designers an enormous amount of freedom.

(It is now certain that the membrane panel switch will replace many other types currently in use, and this is partly because the associated circuitry has fallen in cost with the use of integrated circuits. The switch already described has a limited usefulness because it has to be pressed continuously to maintain a contact. For a positive 'on'/'off' using two pressure points – one for 'on' and one for 'off' – some kind of latching mechanism is needed, and this can now be supplied cheaply electronically. See the *electronic switch, section 4*, for a circuit providing this latching function.)

From an educational point of view, perhaps the most interesting thing about membrane switches is the fact that they can be designed and built by very young children with relative ease. For example, the three layers of the switch described above

19

25 *The Denford ORAC bench training lathe, the first CNC training unit designed for interface with computer systems in common use in education. Control is via the multi-function membrane panel switch seen on the right*

can be produced in thin card lightly tacked together as a sandwich using an adhesive such as 'Pritt Stick'. The conductors on the top and bottom layers can be cut from aluminium kitchen foil and fastened to the card using the same adhesive. And to complete the switch, flexible (stranded) leads are attached to the two lower tracks by small staples to provide a good mechanical as well as electrical joint.

Because the top membrane panel calls for graphics of some kind, the design of the total package represents a rare opportunity to combine some technology and graphics in a single project. In practice, the switch is capable of endless adaptations, and the potential it offers for designing and making in school is quite clear. Successful examples of school work include: push switches for use in signalling, switches for alarm systems and switch panels for programmable games.

The first of these applications is illustrated as a simple circuit in fig. 26, and the outline of a possible design brief is given below. (Fig. 27 reproduces an information sheet prepared by one school for this kind of project.)

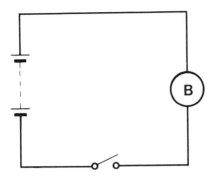

26 Simple continuity circuit with membrane panel represented as a conventional SPST switch

Specimen project brief for a membrane panel switch

There are many ways of telling a household of a visitor's arrival. They include shouting through the letter box, banging on the door, or pressing a button. The last of these is probably the best, and many houses have a small push-button switch to operate the buzzer or bell. Unfortunately, these switches are often small and difficult to see. You are therefore asked to: DESIGN AND MAKE A MEMBRANE PANEL SWITCH FOR OPERATING A BELL OR BUZZER. THE SWITCH IS TO BE FASTENED TO A DOOR OR DOOR FRAME AND BESIDES ATTRACTING THE ATTENTION OF THE VISITOR IT HAS TO FIT IN WITH ITS SURROUNDINGS.

Some points to consider

1 Where is the switch to be used and located? What size should it be? Is it to be serious or fun? would it fit in with what exists?
2 How is the switch to be fixed on? (It might be stuck on using double-sided tape or surrounded, for example, by an acrylic frame screwed on.) Whatever the choice, how are the leads to be taken from the switch?
3 What graphics methods are to be used for the switch top? (Besides felt-tip, rub-down lettering, etc., you might think about the use of found material such as magazine pictures. These can be pasted down and then protected by covering the whole surface with transparent self-adhesive covering film.)
4 How are the conductor tracks to be set out under the graphics

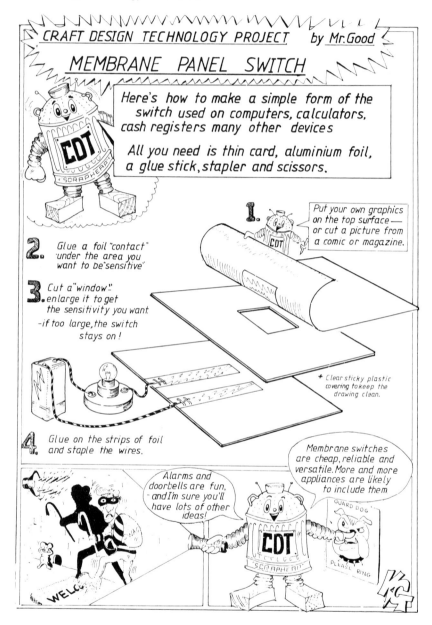

27 *Example of a worksheet prepared for a membrane panel switch project*

panel? (One window in the centre layer gives one point to press; a series of such holes would give a line of possible contact points. But remember – the holes in the centre layer must not be too large otherwise they could allow the top layer to accidentally bridge across the bottom tracks. Remember too that the sensitivity of the switch can be altered by changing the width of the gap between the two bottom tracks.)

Such a project presupposes, of course, the availability of an inexpensive sounder or buzzer. The older electromechanical buzzers and bells have never been very cheap, but the newer solid state units driving a crystal transducer are now often available on the surplus markets, and new devices can be obtained for little cost.[4] Indeed, if the switch is to be the focus of the project – as opposed to a complete system involving a housing for the battery, buzzer, etc. – an LED (light emitting diode) can be incorporated simply to prove the switch.

The membrane panel principle is perhaps already familiar to many readers in the mat switches used for alarm systems. In the example shown in fig. 28 the two outer membrane layers are faced on their inner side with foil and these are normally held apart by an insulating centre layer pierced with a series of windows. When the laminate is assembled, moderate local pressure

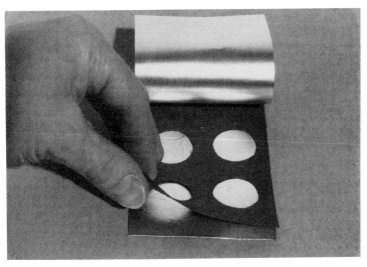

28 *The interior of a commercial mat switch showing pierced felt insulation layer and two outer foils*

practically anywhere on the switch surface will cause bridging of the foils through one or more windows, completing whatever circuit the switch is connected to.

In a typical alarm system, the momentary closing of such a switch sets off a continuous warning sound until this is switched off at some other point in the system. In the case of our very simple buzzer/battery/switch circuit, the sound will continue only as long as the switch is depressed unless, for example, a very simple delay method can be arranged to keep the buzzer running for an appreciable period.

This can in fact be achieved by placing a sizeable capacitor across the battery supply as shown in fig. 29. When the switch is closed, the capacitor charges up more or less instantaneously and the buzzer sounds, but when the switch opens again, the capacitor discharges via the resistor to keep the buzzer running for a period. This time delay depends on the current consumption of buzzer (or LED), the size of the capacitor and the value of the resistor. Many solid state buzzer units have a very low current consumption and will run for a considerable time – e.g., 10 secs – if a 1000µF capacitor is employed. (A standard LED will run for about 5 secs if the supply is 9v and the resistor 2K.) A few seconds of noise or light is not much use in commercial alarm applications, but it is useful to introduce children to what is possible in principle. They can design and produce simple alarm systems using the membrane panel principle and at the same time get a good practical introduction to two essential electronics components.

The delay circuit can also be used in conjunction with other switch types when the alarm is intended, for example, to protect a brief case or bicycle (see *section 4* for details of the construction

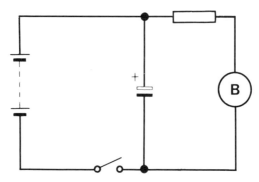

29 *Simple continuity circuit with capacitor delay added*

of appropriate switches). Alternatively, it can be used to good effect in the solution of simple coin-control problems – e.g., money boxes which cause a buzzer to sound or LED to light up when a coin is deposited and momentarily closes a switch. A project involving the design and construction of a complete money box 'system' is a stimulating and challenging one but well within the grasp of younger children. The switch itself can be designed according to a number of principles, some more certain than others.

In practice, a very successful switch consists of two strips of very flexible schim brass (or equivalent material) secured horizontally about 40mm beneath the coin entry slot. They should overlap slightly and remain separated by a small air gap which is momentarily closed when the coin strikes (fig. 30). A good substitute for schim brass is thin plastic sheet or card 'metallised' with aluminium foil. (Note: allowing the coin itself to act as a conductor as it falls between two contact pieces is not usually very satisfactory because of contact resistance arising from the tarnished coin edge.)

30a Switch arrangement for coin-drop operation in a money box
30b Money box incorporating the switch shown in fig. 30a

The mat switch shown in fig. 28 can also be adapted to the design and construction of programmable games. In this application, using the same circuit (with or without delay), the windows in the centre membrane layer are placed in a specific pattern and the complete switch is then overlaid with a graphics panel designed as part of a game for which there are established rules. In fig. 31, for example, the counter in a 'Snakes and Ladders'

25

type game is moved over the square surface of a membrane panel switch according to the score of a thrown dice. When it is placed on certain of the squares, a buzzer sounds – investing the game with an extra dimension of interest. Countless variations on this and other themes are clearly possible.

31 *Membrane panel switch for a 'snakes & ladders' type game – made from paper, card and aluminium foil*

In practice, the area of the switch should be kept as small as possible to reduce the cost of foil and the adhesive needed to apply it. An ideal fixative for larger areas is 3M's 'Spraymount', but this is expensive and it is usually necessary to resort to paste. The important point to remember is to apply the foils without wrinkles or air pockets. If the centre layer is quite thin, any such imperfections in the foil surface may cause localised bridging and spoil the switch. For this reason alone, it is desirable only to tack the membrane layers together so that they can be taken apart easily in fault finding.

Structures

It has already been suggested that designing and making small toys and models can become an object lesson in structures if the range of materials available is sufficiently restricted. In fact, a good deal of introductory work on structures already proceeds on this principle; given an intrinsically weak material such as paper, children might be required to solve the problem of supporting a heavy load – e.g., *using only a sheet of A4 size paper, support a house brick 50mm above a flat surface.* This kind of activity is designed to encourage an intuitive grasp of how structures behave and, ideally, it ought to establish some kind of conceptual foundation capable of later development. The discussion which follows examines two ways of making quite difficult ideas intelligible to younger children through demonstrations and practical involvement in designing, making and evaluating.

Soft foam modelling

There are several good commercial systems for modelling structures involving the use of tension and compression elements (e.g. springs) which visually exaggerate the effects of applying loads and forces. These can be excellent aids for teaching about structures especially during what one might describe as the pre-quantitative phase of learning.

But equally effective and much cheaper as a teaching medium is the soft foam used so widely in upholstery work, etc. This can be cut into pieces representing a variety of structural elements, and these can be assembled on a board to model quite complex structures. When loaded, the deformation of such 'structures' corresponds in principle to what would be expected from more rigid materials but the degree of movement is, of course, exaggerated.

This use of foam is not new; many physics specialists, for example, have made *ad hoc* use of the material in their teaching – but without realising perhaps its full potential as a means of comprehensively illustrating structural phenomena in tangible and exciting ways. A great deal can be taught using this system, though limited space here allows only a couple of examples. All that is required is a block of foam (best cut with a bandsaw), a display board angled at about 30° and some black spray paint with paper for masking.

Fig. 32 illustrates an obvious first experiment using the foam system. A simple rectangular section 'beam' has been sprayed on one surface through a paper mask to give a series of parallel reference lines. It is placed between two supports on the board and loaded with weights so that the compression and tension effects are shown in the way the grid lines open and close

32 The behaviour of a simple beam under load demonstrated using soft foam

respectively. Obvious questions arise from this simple demonstration; for example, how can the beam be reinforced in the most economical way? After directed discussion, a length of thread can be attached along the underside of the foam fastened with tape at either end. When the deflection of this modified beam is compared with the first, it establishes quite convincingly that the reinforcement works and leads into a consideration of why it does so.

Similarly, the system can be used to illustrate how cantilever construction 'works' or to demonstrate important characteristics of entire building systems (fig. 33). And it is even possible to explain by modelling something as esoteric as the necessity for flying buttresses in medieval cathedrals, an achievement which – if it does nothing else – certainly highlights the versatility of a modelling system that can be adapted by the CDT specialist to meet almost any contingency.

Straw modelling

'Artstraws' have been in use for some time as a structural modelling material, but it is often suggested that they come a poor second to balsa wood. This is contrary to the experience of many CDT specialists who find in the use of these straws a material more true to scale than balsa and certainly more affordable. The construction of models using straws is very straightforward, and their characteristics can be compared semi-quantitatively with very simple improvised apparatus.

Figs 34-37 illustrate stages in the construction of one of the

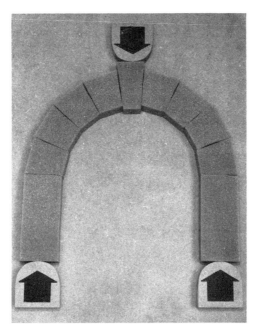

33 Examination of an arch under load using soft foam components

ubiquitous trusses using a type of construction familiar to most model makers. The truss outline is drawn out full size on paper and lengths of straws are secured temporarily over this with masking tape tabs (prior to which wax is rubbed over the joint areas). Using a syringe or similar dispenser, each joint is then 'spot welded' with the application of a blob of white glue (e.g., 'Resin W'). When the glue has set, the complete truss can be removed from the paper for direct examination or incorporation into a larger structure.

Any modelling of this kind makes little sense, of course, unless it is part of a scheme that seeks to compare, understand and improve the product. Ideally, what is required is an objective method of comparing solutions to problems, and this is provided by the very simple piece of test apparatus shown in fig. 38 – being used here for investigative work taking place at third-year level. It uses the familiar steelyard principle to apply a load via a weighted arm, connected to the truss close to the arm's pivot point. The great advantage of this arrangement is its adaptability for compressive as well as tensile loading experiments (fig. 39),

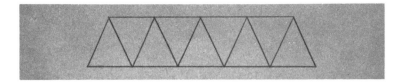

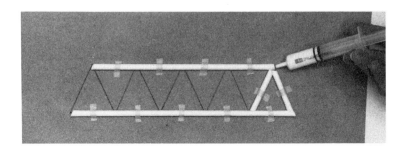

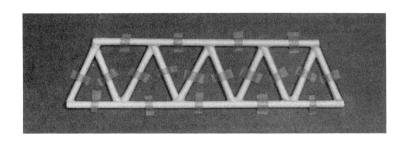

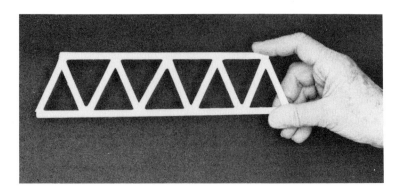

34/37 Stages in the construction of a simple truss using 'Artstraws'

and the fact that it allows deflection of a model to be measured against the load applied. Quite small deflections are amplified considerably when the measuring scale is placed towards the end of the arm, making it quite easy to record incremental movements against increased loads and to plot simple load/deflection curves.

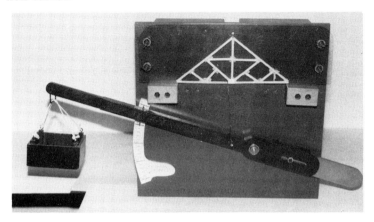

38 *A simple piece of test apparatus used to compare truss forms*

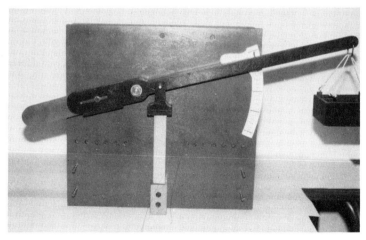

39 *The apparatus seen in fig. 38 (above) used for compressive testing*

The experiment in progress shown in fig. 38 examines the load-bearing characteristics of a particular type of truss. Although there are distinct problems in looking at and comparing two-

31

dimensional structures in this way, many important principles can be established by noting carefully how the model fails. In the example shown, collapse will ultimately almost certainly arise from a compressive failure at the top of the model. For most children this is at first contrary to common sense, but they quickly learn to think about modifying and reinforcing such models in appropriate ways. They are also quick to begin asking questions about the validity of the experimental method itself: how should the truss be supported and where should the load be applied?

Notwithstanding the experimental problems, the apparatus can be of immense help in clarifying quite abstract ideas such as 'efficiency'. In designing a structure to span a specified gap, for example, the most efficient solution can be said to be that which supports the greatest load for the least expenditure of material. If the maximum load supported prior to failure is divided by the total length of all the straws used, the resulting number (in arbitrary units) can be taken as a measure of comparative performance; i.e., the higher the figure, the more efficient the structure.

Of course, considerable caution is called for in interpreting these results and making judgments. But the possible pitfalls seem to be heavily outweighed by the opportunities for motivating children and teaching them effectively through practical involvement.

Mechanisms

Technology has already made inroads into lower school CDT, and nowhere is this more apparent, perhaps, than in some of the excellent work involving mechanical movements – animated figures, faces, cam-operated toys, and many more besides. Much of this work is well documented, and I have very little to add here. What follows is really offered as an extended *footnote* to existing initiatives; in particular, it draws attention to a wealth of neglected source material, and it seeks to illustrate one or two variations on the increasingly important and popular theme of cardboard engineering.

The last hundred years has seen a surprising number of publications exclusively concerned with the encyclopaedic classification and illustration of mechanical devices for control, power transmission, etc. Although systems of classification may change, the basic material seems to endure remarkably well – largely because the mechanical principles themselves are enduring. Many of these older texts are a goldmine, especially for the CDT specialist who is looking for clear and concise factual information and who may be seeking inspiration for new project work. Although earlier material remains scarce, twentieth-century books can usually be obtained without too much trouble through the excellent inter-library loan scheme. In response to many requests during in-service courses, a short list of such publications is given in the notes at the end of this section together with details of one or two contemporary publications worth consulting.[5]

If older books can still be drawn on as a valuable resource, there are those too which are quite inspiring in the pedagogical sense. Many early twentieth-century technical publications were designed to help readers educate themselves through the manipulation of sectional overlay illustrations or articulated card models. Fig. 40, for example, shows a 'working' diesel engine (circa 1900) capable of showing ten distinct stages in its cycle of operation as the interconnected parts are moved.

It is hardly surprising that this kind of material is now enjoying something of a renaissance with 'pop-up' amusement books coming out by the dozen and serious educational material beginning to appear once more.[6] The sheer technical brilliance of some of these publications is alluring enough, but they have a special interest for the CDT specialist because they point to so many ways in which children – especially younger ones – can get involved in designing and making things that work.

The same methods used to articulate the model seen in fig. 41, and employed in some of the newer publications, can be applied to various types of sheet plastic as well as card or paper. But the great advantage of plastics is their ultimate durability and the precision with which linkages, etc. can be made to operate. This means that it is not only possible to solve problems through modelling, but to design and create products such as drawing instruments which exemplify important mechanical principles *and* can be put to work.

The pantograph copier shown in fig. 41 has been assembled from 0.5mm thick polystyrene sheet with the sections pivoting very accurately around the 'Velos' eyelets normally employed for reinforcing punched holes in paper and card.[7] A special combined punch and eyelet-closing tool is obtainable with the eyelets at good stationery shops, but it has a rather narrow throat and its use over larger areas is therefore limited. As an alternative, holes for the eyelets can be drilled conventionally

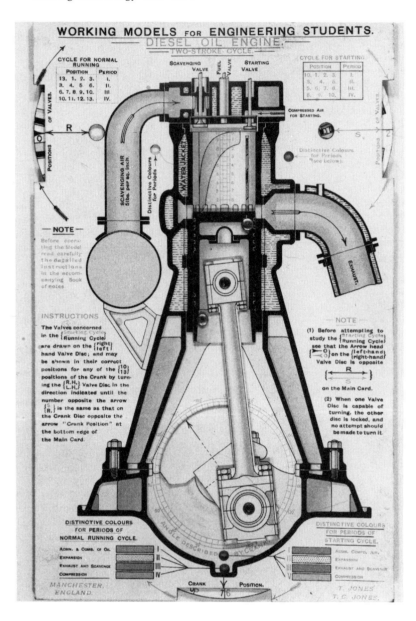

40 *Articulating educational model of a diesel engine*

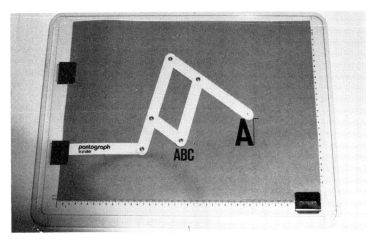

41 Pantograph constructed from strips of polystyrene seen in use here on a commercial drawing pallet

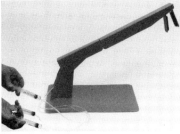

42a Lazy tongs mechanism constructed in corrugated plastic material

42b Remote control 'robotic' arm using syringe hydraulics. This device has been constructed exclusively in corrugated plastic sheet bonded with hot-melt glue

and a ball-nosed punch made up for opening them out once in position.

The eyelet system is exceedingly useful too for other varieties of mechanical modelling. Fig. 42a shows how the eyelets are used to create a fully articulating lazy tongs mechanism. But the distinctive thing here is the sheet material itself: the polypropylene equivalent of corrugated cardboard now so familiar as estate agents' signs, carrying cases for papers and books, and many other applications where a developed sheet can be folded and clipped together. It is an exceedingly cheap, colourful, and easily worked material with great potential as a modelling medium for younger children. And, as fig. 42b indicates, it is also ripe for application in more sophisticated modelling and proto-typing. The 'robot' arm illustrated is constructed entirely in the fluted polypropylene sheet using hot-melt glue for butt joining, plastic pins for pivots and the material itself folded and creased for hinges. This is clearly a material with a future in CDT.[8]

Notes

1 Supplier: Downswood Products Ltd, Park Lane, Knebworth, Herts, SG3 6PJ.
2 Supplier: C.F. Anderson, 7/9, Islington Green, London N1. (Tel. no. 01 226 1212) (The above firm has other branches in different parts of the country.)
3 Supplier: Arterial Medical Services Ltd, 313, Chase Road, Southgate, London N14 6JH. (Tel. no. 01 882 4434) (Contact firm for details of minimum quantities.)
4 E.g., contact local surplus firm(s), local surplus buying agencies, SATRO's etc.
5 See, for example, the following (listed in chronological order):

'Mechanical movements', *Spons' Dictionary of Engineering*, Div. 7, 1873, pp. 2421-2452, London: E. & F.N. Spon.
This is a very rich source book for those interested in the history of technology. Hundreds of mechanical movements are illustrated in the section referred to, and this is especially noteworthy since many of the plates are re-used in thesaurus-type publications a century later!
T.W. Barber, *The Engineers' Sketch Book of Mechanical Movements, Appliances, Contrivances, & Details*, London: E. & F.N. Spon, 1890.
G.D. Hiscox, *Mechanical Movements, Powers and Devices*, London: Constable, 1919.
H. Herkimer, *Engineers' Illustrated Thesaurus*, New York: Chemical Publishing, 1952.
D.C. Greenwood (ed.), *Mechanical Details for Product Design*, McGraw-Hill, 1964.
G.N. Sandor and A.G. Erdman, *Mechanism Design – Analysis and Synthesis*, Prentice-Hall, Englewood Cliffs, 1984. A recently published in-depth study which examines, among other things, CAD techniques applied to the design of mechanisms.

The following publication remains a useful as well as fascinating resource for more advanced work:

F.D. Jones (ed.), *Ingenious Mechanisms for Designers and Inventors*, New York: Industrial Press, 1935.

Getting technology into CDT

This is the first part of a multi-volume work added to and published over a number of years. Although microelectronics and the microprocessor have done away with our dependency on many of the mechanical control and programming systems illustrated in this book (e.g., compare several recent generations of sewing machines), many of the principles discussed are as relevant today as fifty years ago.

6 See, for example: P. Moore, *The Space Shuttle Action Book*, Aurum Press, 1983. R. Marshall and J. Bradley, *The Car (Watch it Work)*, Sadie Fields Productions, 1984.
7 Supplier of polystyrene sheet: Amari Plastics PLC, 2, Cumberland Avenue, Park Royal, London NW10 7RL.
8 Supplier of corrugated plastics sheet: Robert Horne, Product Promotions, Huntsman House, Unit 3, Bermondsey Trading Estate, Rotherhythe Road, London SE16 3LW. Supplies of Artstraws can be obtained from: Nottingham Educational Supplies Ltd, 17, Ludlow Hill Road, Melton Road, West Bridgeford, Nottingham NG2 6HD.

Section 2

Manufacturing technology

It is not the intention here to examine a wide range of manufacturing methods, but to show how several of those perhaps less familiar to CDT specialists but important in industry – including some still regarded as exotic – can be demonstrated in school and exploited in designing and making. In some cases, this will involve the construction of simple equipment, but it will become clear that the requirements in cost and time are minimal, and it is suggested that the results will amply reward the effort. For those interested in further detailed information on any of the techniques under discussion, good bibliographies can be found in books referred to in the notes section.[1]

Metals: some cold working techniques

Hydraulic forming

Hydraulic forming is now a well-established method of forming metals in sheet or tubular form by means of fluid under high pressure. In the manufacture of specialised exhaust systems, for example, a length of steel tubing can be plugged at both ends and 'pumped up' with hydraulic fluid. Simple restraining collars and/or dies placed along its length control expansion locally to produce a profiled tube which might be difficult, if not impossible, to manufacture in one piece by another method. (Conversely, the tubing can be reduced in diameter by enclosing it in a heavy steel cylinder and introducing fluid under pressure in the space between the two.) Another interesting application is to be found in musical instrument making where brass tubing which has to be formed into very tight bends is first pressed flat, formed into the required curve in that condition, and then pumped up again within a suitable die.

The principle can be extended, of course, to forming sheet metal providing that this can be clamped down and sealed to allow hydraulic fluid to act under pressure on one side – pushing the sheet into a suitable die on the other. In practice, the equipment used is a high-pressure pump feeding fluid into a thick-walled reservoir over which the sheet is held. The workpiece then becomes, in effect, the weakest part of a pressurised enclosure.

This is clearly a fascinating and useful technique for metal deformation, but one requiring specialised and often very expensive equipment for industrial application. Very good seals are required to contain fluid under high pressures, and although one could envisage a simple form of pump, it would be quite time-consuming to make (or convert) one.

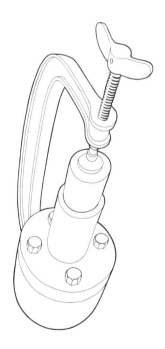

43 *General exterior view of a 'Plasticine' hydraulic former employing a 'G' cramp to load the piston*

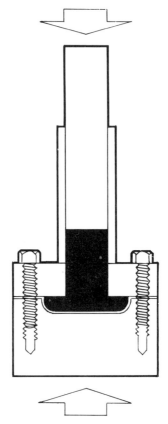

44 Sectional view of the hydraulic former showing workpiece deformed under the pressurised 'Plasticine'

For small-scale work, both of these problems can be avoided by using 'Plasticine' as the hydraulic fluid with a very simple piston/cylinder arrangement acting as the pump. Figs 43/44 illustrate such a system in which the sheet to be deformed is trapped around its edge between the flange attached to a cylinder and a die of equivalent diameter – all of which are firmly held together either by bolts or the use of some form of external clamp. When the piston is forced down, the 'Plasticine' is pressurised forcing the sheet into the die, and if an open ring were substituted for the die shown, the sheet would simply expand out until a combination of thinning and strain hardening caused it to rupture.

Depending on the overall size of the device, the piston can be forced down using a 'G' cramp, fly press, or even a large bench vice. Whatever system is used, the ultimate fluid pressure obtainable can be increased by reducing the diameter of piston and cylinder. However, this correspondingly reduces the amount of 'Plasticine', and it may therefore be necessary to complete a forming operation in several stages, topping up each time with more 'Plasticine' and perhaps taking the opportunity each time to anneal the sheet.

The construction of a complete hydraulic former using 'Plasticine' can be very straightforward indeed. Fig. 45 illustrates one such device produced entirely in scrap materials, and presented here in order to show how little is required in materials and labour (construction time – approximately 1½ hours). This particular example was produced to form a number of identical silver dishes (diameter 80mm) from circular blanks (fig. 46). The die is a piece of aluminium turned out to the profile of the dish, and the pump a length of steel gas piping force fitted into a turned steel flange. The piston is simply a length of steel turned to provide a reasonably good fit in the cylinder. Actual dimensions of this prototype are: cylinder bore – 35mm, length of stroke – 70mm. It will therefore displace approximately 60cc of 'fluid' in one cycle of operation.

Although seamless tubing is ideal for the cylinder, a seamed tube – even one with a relatively rough wall – can be used. There will naturally be a tendency for 'Plasticine' to leak backwards between the piston and cylinder wall under these conditions, but the 'Plasticine' is an extremely viscous fluid and effects an excellent seal as it leaks back in this way. A penalty is paid, though, since the piston becomes increasingly difficult to withdraw from the cylinder after the forming operation. If nylon is available, a turned piston with a feather edge seal obviates this problem to some extent. (It is also worth noting that a piece of thin paper placed between workpiece and 'Plasticine' prevents the latter from sticking to the surface.)

Fig. 47 shows a completely self-contained version of the former, identical in principle to the first, but using a hydraulic car jack to apply pressure to the piston. Component parts of this former are shown in fig. 48, and it can be seen that the cylinder flange is bolted to the circular die – which has a cavity bored out in order to accommodate different die inserts. These can in fact be manufactured from a variety of materials including wood and plastics, but precisely what material is employed naturally depends on the type and gauge of sheet metal to be formed and the fluid pressures involved.

41

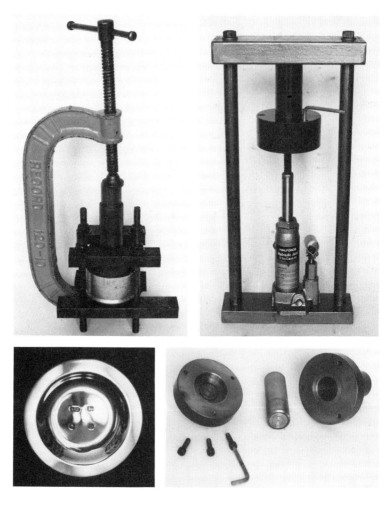

45 (top left) Prototype hydraulic former constructed from scrap materials. The two halves of the former are held together here with two pairs of clamping bars

46 (above left) Hydraulically formed silver dish – using the 'Plasticine' system

47 (top right) An actual version of the former seen in fig. 43 (above) loaded here with a small hydraulic car jack contained in a simple steel frame

48 (above right) Separated components of the bolt-together hydraulic former used in the frame seen in fig. 47

The thought of using softer die materials is a very attractive one since children can design *and* produce die inserts for quite complex pressings entirely by hand. Where very thin sheet metal or foil is to be used in low-relief work, cardboard – or even paper – can be cut out and laminated to form the insert; examples of pressings produced in this way include badges, button coverings, model wheel hub caps, etc.

It is possible also to construct metal dies without the need for any machining. Perhaps the quickest method is to model the required relief form in wax, plaster or 'Plasticine' itself and, after building a thin retaining wall around it, pour over a quantity of fusible alloy such as Wood's metal (MP:70°C).[2] At this low temperature, no harm comes to the modelling material which is afterwards picked off the metal casting to leave a hard and quite durable die. When necessary, the alloy can be recovered by re-melting. (WARNING: Wood's metal is an alloy of tin, lead, cadmium and bismuth, and is therefore potentially toxic. Ideally, it should be melted in a vessel surrounded by water (i.e. a water bath) or melted under water, which is decanted off prior to pouring. The material should also be handled with gloves since rub-off occurs so easily.)

Using the equipment and methods described, there can be few easier ways of getting from an idea to a finished product without the necessity for complex machining. By its very nature, hydraulic forming does away with the need for exactly toleranced intermatching die sets. And the version of hydraulic forming discussed here simplifies the process even further and gives even younger children an opportunity to learn about it in a meaningful way and to realise ideas that might not otherwise have been possible.

It hardly needs adding, perhaps, that the design and construction of the hydraulic former itself presents a most interesting challenge to the older pupil or student.

Cryogenic forming

Of all the more exotic metal-forming techniques, cryogenic forming is perhaps the most interesting as well as the most straightforward one to demonstrate. It is well known that a considerable expansion takes place when water passes into its solid phase: ice. This expansion – roughly 10% by volume – can cause appreciable movement and damage when water is enclosed – for example, in domestic plumbing systems or water-cooled engines.

Michael Faraday, during several of his celebrated Royal Institution lectures, gave dramatic demonstrations of the phenomenon by placing water-filled flasks of cast iron into a freezing mixture of salt and ice, with explosive results when the contents froze. Similar nineteenth-century experiments included filling up cannon balls with water, sealing them and then freezing. The results were invariably the same – violent expulsion of the sealing plug or total rupture of the cast iron shell.

The mechanical forces involved here are clearly enormous, and adequately controlling them is the key to very impressive feats of metal deformation. Fig. 49 illustrates a device for cryogenic forming of sheet metal or plate. The workpiece is sandwiched between the two halves of a thick-walled steel cylinder which are held together by long bolts passing through steel plates covering the cylinder ends. The bottom half of the cylinder under the workpiece is filled with water, and a suitable die is fitted into the top half. Once tightly clamped together – a good metal-to-metal seal is sufficient between workpiece and cylinder – the whole device is placed in a domestic freezer.

The results are invariably interesting, especially when it is realised that the device will take any opportunity to destroy itself. If the workpiece is the weakest part of the enclosed system, it will

49 Simple cryogenic forming apparatus

50 (right) Elastomeric forming of sheet metal using a 20-ton capacity bottle jack to load the forming tool seen trapped at the top of the steel frame

deform into the die as required. But if too much water is used relative to the volume of the die cavity, either the cylinder, the clamping bolts or both will also be deformed as the expansion continues. It is necessary, therefore, either to measure the volume of water used very carefully and/or build in a safety valve so that excessive expansion is harmlessly relieved after deformation of the sheet.

The volume of water can be adjusted by the simple expedient of displacement with, say, ball bearings. But, assuming that it might not always be possible to gauge the volume of water precisely, an alternative solution would be to place a block of predictably crushable material such as wood behind the die. In practice, this and other materials have been used – including rigid foam, collapsing metal tubes, and even bricks.

Cryogenic forming is employed in a number of highly specialised metal-forming operations, but as a commercial technique it is not all that common. However, it takes very little to prepare a cryogenic forming cell either for demonstration purposes or to do some useful work when heavy-duty mechanical or hydraulic press equipment is not available. A forming cell is easy to construct, and there are no special sealing problems since the ice forms initially against the metal surfaces and effects its own high-pressure seal. (But it is convenient to bond the bottom half of the cylinder to the base plate to prevent water leaking out in the first place.) Successful operation of the former really depends on making it strong enough to resist the forces involved – and taking care not to incorporate *cast* materials in its construction.

Elastomeric forming

Elastomeric and hydraulic forming appear very similar in principles if one thinks of an elastic rubber block acting under pressure on the workpiece rather like a fluid. In a typical elastomeric sheet metal forming operation, the sheet – for example, a plate of aluminium which is to be lightly ribbed – is sandwiched between a steel die and a rubber block backed onto a platen. When pressure is applied, the rubber is squeezed into the die with the aluminium correspondingly deformed in between the two.

Different types of rubber can be used in a variety of ways for forming metals, again avoiding the need for expensive inter-matched dies.[3] Indeed, where products incorporate re-entrant

45

sections, etc., elastomeric forming is very often the only realistic answer to a production problem. For example, a deep-drawn container which requires expanding out into a barrel form can be shaped by compressing a cylindrical slug of rubber inside it. Under such pressure, the container is expanded into an appropriate die, and when the load is removed the rubber returns to its original condition for withdrawal.

*51 Pressing in sheet aluminium of a small wheel hub cap seen here together with the die in which it was formed (*right*). The rubber pad is not shown*

Elastomeric forming is a technique that can be demonstrated in several ways without making up special equipment. For example, a simple die for low-relief work can be produced using Wood's metal (as described above), plastics, wood – or just thick card. Commercial elastomers suitable for this work are expensive, but discarded pieces of thick rubber sheeting or materials such as hard 'Vinamould' can be used instead providing they are resistant enough for a particular operation. Fig. 50 shows an arrangement for elastomeric forming that utilises the same frame and car jack illustrated in fig. 47. It is being used here for the manufacture of small wheel hub caps with a turned aluminium plate for the die and layers of tyre inner tube as the rubber block (fig. 51). Although this equipment takes very little time to assemble, there are even easier alternatives. A 'tabletop' demonstration can be put together using the same die and rubber just described together with a pair of steel plates and 'G' cramps. Even a coin or two pushed into aluminium foil against a pencil eraser can produce a convincing demonstration.

Metal-forming: some electrochemical techniques

Electrochemistry applied both to metal decoration and metal-forming has a history stretching back into the first half of the nineteenth century. The oldest extant electroplated object from this period is a small silver goblet described in contemporary records as 'a very small *galvanic* goblet', and the hallmark dates it at 1814. Since that time, an extraordinary range of techniques has been developed not only in relation to decorative metalwork but more recently in the context of precision engineering. Electrochemical machining (ECM) is now a well-established technology, and one of rapidly growing importance.

In view of its long history and current importance, it seems all the more strange that electroplating and associated techniques should have been so neglected in school and college work. An obvious explanation is the highly toxic nature of many of the chemicals used, and one can clearly say that the commonly used cyanide compounds and very toxic etchants are quite out of the question for school use. However, if certain limitations are accepted, relatively safe solutions can be used in demonstrating and utilising a wide range of electrochemical techniques. (See SAFETY NOTES at the end of this section.)[4]

Electroplating

Electroplating involves the deposition of metal by electrochemical action in a bath containing an electrolyte, a solution consisting in its most basic form of dissolved salts of the metal to be deposited. Fig. 52 shows the arrangement for a simple electroplating bath; when current flows in the circuit completed by the partially conductive electrolyte, metal will go into solution from the *anode* (metal to be plated out onto the workpiece) and eventually deposit out on the *cathode* (workpiece). The electrolyte around the anode will tend to become locally enriched, and around the cathode locally depleted; for this reason – together with the fact that gas is liberated at both anode and cathode – the solution and/or workpiece would normally be agitated.

Most commercial plating electrolytes involve the use of cyanide, and it is very difficult to obtain good results in operations such as silver plating without it. One electrolyte, however, which is reasonably safe is the simple copper/acid bath used for plating over certain non-ferrous metals or building up copper. Commercially, its potential is quite limited, but its use

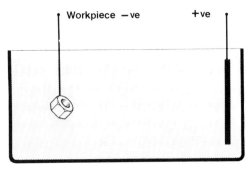

Workpiece — ve   + ve

52 *General arrangement of a simple electroplating bath*

does open up an enormous range of possibilities for CDT work.[5] It should also be noted that successful results with this bath on a small scale can be obtained without satisfying all of the normally exacting commercial conditions which can make the operation both difficult and expensive. If the procedural guidelines set out below are followed, the results should be more than adequate.

The electrolyte for copper plating can be made up in the following proportions:

Copper sulphate	220 grams/litre (of water)
Alum	12 grams/litre
Sulphuric acid (pure)	26 grams/litre
(Sp. gravity 1.84)	

The usual precautions should be observed in making up the electrolyte – i.e. acid is added slowly to water, and the solution should be clearly marked with warning signs. In setting up and using the bath, the following points are offered as guidance in note form:

1 The bath can be operated within a wide range of room temperatures and, contrary to the normal textbook insistence on distilled/de-ionised water, tapwater can be used – remembering, however, that mineral content will vary from area to area.

2 A medium size battery within the voltage range 1.5 to 4.5 volts can be used to operate the bath – e.g. HP2 size cells used singly or in series. No instrumentation is necessary for the small and uncomplicated work envisaged here.

3 A pure copper anode is desirable, but not strictly necessary. Commercial copper sheet will prove quite adequate.

4 Although it is desirable to clean the surface of the workpiece

chemically prior to plating, mechanical cleaning with metal polish will give good results providing all traces of polish and dirt are removed.

5   After it is made up, or following a period of use, the bath may acquire an inhibiting surface film which will wrap itself around the workpiece as this is lowered into the electrolyte. Before use, therefore, the surface of the bath should be 'wiped' slowly with the edge of a piece of paper so that the film clings to it and is removed.

6   A stout glass or other acid-resistant container should be found to hold the electrolyte. In a commercial operation both the shape and the positioning of the anode(s) can be important, but here a single copper sheet bent over and hooked to the top edge of the container will suffice. As a very rough 'craft' guide, allow at least 50 cm^2 of contact area per litre of electrolyte.

7   The workpiece (cathode) can be held in the bath with a pair of brass tongs wired to the negative side of the battery, but this naturally entails altering the place of contact with the jaws periodically to avoid masking small areas of the work. In the absence of stirring equipment, the work can be agitated quite easily in the bath with the tongs.

A typical sequence of operations in plating a small sampler of brass with copper would be as follows. After preparation, which might include buffing, the sampler is cleaned with metal polish and wiped clean on dry cloth or tissue paper. As a rule of thumb, the surface will be clean enough when no further black deposit appears on the cloth or paper. Unlike the action of some plating baths, though, this one will not improve the surface of the work. If a flash coating of copper is deposited, the surface will remain shiny but as further material is built up the appearance alters to a dull salmon pink – which can be polished as a secondary operation.

When ready, the workpiece is gripped firmly in the tongs with minimal area of contact, and then immersed in the bath and agitated. It should be removed for periodic inspection and turned in the tongs to avoid masking by the jaws. If nothing happens at this point, it is almost certainly due to an electrical fault, and all battery connections etc. should be checked. If the result is thin and patchy plating, the reasons can be several – poor surface preparation, lack of agitation, inhibition caused by film pick-up from the bath surface, or all of these. A dark brown deposit over the entire surface or – more typically – at the edges and corners, indicates a different problem: too high a voltage. This brown granular deposit, known as *burning*, means that the copper is plating out too rapidly. With a 3 volt supply, a flash deposit of copper will take just a few seconds to appear on a flat sampler having a total surface area of some 20 square centimetres. A reasonably hard-wearing film would take about 30 seconds to build up.

Unfortunately, steel cannot be plated directly with this bath, and brass is one of only a few metals which can. But apart from its usefulness for demonstrating an important process, some interesting work is possible even within these limits. For example, dry transfer rub-down lettering systems such as 'Letraset' are unaffected by the electrolyte and can be used as a stopping-out medium. If a small piece of brass sheet is stopped out in this way and copper plated, a very sharply defined brass-against-copper image appears when the lettering is removed. And it is not without interest that this principle is applied commercially in the production of nameplates, labels, etc.

Electroforming

The thickness of the electrodeposited film on the workpiece depends on a number of factors, but given time the entire anode will eventually erode away and re-appear on the cathode. The deliberate growth of thick metal shells by protracted plating has long been an important means of forming metal products, and the technique is still widely applied in printing, for example, and decorative metalworking – e.g. the Prince of Wales' crown. Electroforming is being used increasingly for the manufacture of small metal products, especially where production runs are small and high-dimensional accuracy and good surface finish essential.

In a typical commercial electroforming operation, the 'tooling' consists of a *mandrel* upon which the metal shell is grown and then removed. Mandrel materials include metals such as stainless steel and aluminium, fusible alloys (which permit the mandrel to be melted out after use), and non-conductors such as plastics and glass which require to be given a conductive surface. For these non-conductors, there are several possible surface treatments such as vacuum deposition of thin metal films and spray application of silver-loaded paints. Unfortunately, the first of these requires highly specialised equipment and the second can be very expensive. One traditional method, however, which remains both effective and cheap is the application of graphite either in powder form or as a colloidal dispersion (e.g. 'Aquadag').

It is normal commercial practice to grow metal shells on a well-finished *negative* mandrel so that when the product is removed it has a correspondingly good exterior surface finish. It is also possible, though, to do the opposite and grow the shell entirely around a positive mandrel – accepting the progressive loss of surface detail and finish as the shell thickens. This is not an ideal method for other than certain kinds of decorative metalwork, but this kind of electroforming is possible in school. A proven procedure for manufacturing small metal products using the copper bath already described is as follows:

1 Model the object to be electroformed in 'Plasticine' or an equivalent medium.
2 Produce a mould from the model by applying several layers of latex rubber or any suitable cold-setting mould-making material. (For simple demonstration purposes, these two stages of preparation can be side stepped by using commercially produced moulds – e.g. those supplied for moulding chess pieces.)
3 Melt a quantity of paraffin wax and stir in as much powdered graphite as possible, consistent with the mixture remaining just fluid enough to pour at a temperature which will not injure the mould.
4 Pour the graphited wax mixture into the mould, but before setting takes place introduce a conductor (e.g., a 2″ brass wood screw) into the open end of the mould so that when solidification has taken place it can be used to suspend the wax mandrel in the electrolyte and provide a connection to the current source.
5 Holding the wax mandrel by the conductor, dust the surface with powdered graphite and thoroughly polish with a soft brush until uniformly shiny. It should be noted that the success of the initial plating depends almost entirely on the amount of time invested in this operation.
6 Using the bath described above, connect the mandrel as the cathode and commence plating. After about 15 seconds, the mandrel should be removed and inspected; if any part remains unplated, dry in a forced draught and go over those areas again with graphite. After blowing away excess powder, begin plating and inspect again to determine if the process needs repeating.

Depending upon conditions, a shell of some 0.1mm can be built up within 15 minutes. A more rigid self-supporting shell would obviously take longer, but since graphite and wax are quite cheap, there seems little point in attempting to recover them by melting out unless it is important to end up with a hollow shell. A very thin shell of copper permanently supported by the wax inside is very durable and can be polished by all conventional means. If the wax is to be melted out, however, stearene should be added to the mixture in the first place to minimise the problem of expansion when the wax is re-heated in the shell. The expansion coefficient of paraffin wax is very high, and the shell can easily be cracked open.

Because the amount of metal deposited is proportional to the current passed, it is worth considering a low-voltage power pack for this work rather than expensive batteries. It must be stressed, though, that only approved commercial equipment should be used, and that any supply unit should be kept clear of the bath to eliminate possible dangers from electrolyte spillage. Because of the length of time involved in this operation, it is also worth considering the addition of mechanical stirring equipment.

The advantage of the graphite/wax technique is that it provides a mandrel easily moulded to shape *and* partially conductive; it thus requires less surface treatment with powdered graphite than would otherwise be the case. Alternatively, the use of colloidal graphite, in the form of the commercial product 'Aquadag', makes it quite easy to produce a good conductive film on a non-conducting material. The only real problem might be obtaining a smooth surface if a brush is used to apply the 'Aquadag'. Nevertheless, it is well worth experimenting with this form of graphite – remembering to attach a piece of 'jigging wire' for the anode connection to the mandrel first of all.

Similarly, taking great care in melting and handling the material, it is well worth producing a simple mandrel in Wood's metal and forming a shell over this. It plates very readily, and is easily melted out under water for re-use.

Electrochemical machining (ECM)

In any electroplating operation, the anode will be sacrificed as the cathode grows in size unless an inert anode is employed and the electrolyte replenished with fresh salts. If attention is therefore directed to the anode disappearing rather than the cathode growing, it becomes clear that a means exists for deliberately removing metal. If, for example, a series of small square holes are required in thin-gauge copper sheet, electrochemical removal offers a convenient and distortion-free method. The material to be retained is stopped out on both sides of the sheet which is then exposed in a suitable bath as the anode (e.g., the copper/acid bath described above). The exposed areas are eventually eaten through to leave only the desired stopped-out part of the workpiece. In principle, this is the process used in several commercial operations including, for example, the production of pierced foils for electric shaver heads and micro-mesh filters.

In recent years, the principle of electrochemical removal has been exploited to shape hard alloy steels with great precision and control. Purpose-built equipment is now available for performing a wide range of 'machining' operations which would otherwise be virtually impossible by conventional mechanical means.

A simple arrangement for demonstrating the removal of steel in this way involves nothing more than the bath described above with a brine solution substituted as the electrolyte. This is made up with ordinary table salt dissolved in hot water to get as much as possible into solution. If a sample of steel (e.g. HSS), connected as the anode, is immersed in the bath, it will be eaten away at a rate determined by the current passing. There is no plating out onto the cathode under these conditions; the stripped material combines with the electrolyte which must eventually be replaced. This means that the cathode can be practically any conductor such as stainless steel, copper or lead.

In order to expedite the electrochemical action, it may be tempting to use something like a battery charger to supply current. This must be avoided at all costs since few, if any, commercial chargers will meet the exacting safety requirements for safe work in school. Approved equipment should be used, and it should also be noted that this work must only be carried out on a small scale and in a suitable place since the operation of the bath gives rise to toxic fumes. The spent solution should be safely disposed of.

It is possible, at little further expense, to arrange a more specific 'machining' process by pumping the brine against the anode workpiece through a microbore tube (connected as the cathode) and thus confining the electrolytic action to a very small area. Since material is removed only locally, this amounts to a drilling operation if the tube is gradually fed into the workpiece. And it can be a remarkably effective one on a material such as high speed steel.

The practical requirements for a simple ECM drill are a short length of microbore tubing, either found or manufactured (e.g., a brass rod turned down and drilled out to give 0.5mm/1mm bore); a standard car windscreen washer pump, a length of PVC tubing, and found or manufactured fittings to connect the PVC tubing to the side of the bath and microbore tube.

Fig. 53 illustrates the complete system, with the electrolyte being drained from the bottom of the bath to be pumped under pressure against the work. The anode and cathode connections are made, respectively, with a large crocodile clip attaching to the workpiece and a solder join to the microbore tube. The same power supply used to operate the pump will also serve for the electrolytic action providing that a relatively high current (e.g. 5 amps) can be drawn from it.

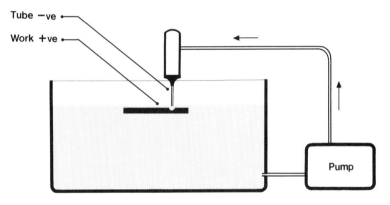

53 General arrangement for an electrochemical drill

The workpiece should be supported just above the level of the electrolyte, and the microbore tube brought to within at least 0.5mm from its surface. When the system is operating correctly, the brine flowing from between the tip of the tube and the workpiece will appear cloudy, and as a depression begins to form, the electrolyte will tend to splash up. Unfortunately, the hole will end up roughly conical in cross section unless the sides of the microbore tube are insulated right down to the tip. But this is not too much of a problem if the occasional demonstration is confined to drilling thin-gauge materials (e.g. hacksaw blades).

Manipulating the microbore tube freehand is quite difficult, and usually results in a lot of short-circuiting. If the drill is to be used effectively, a mechanical system capable of providing a slow and precise rate of feed is required. Just as the design and construction of hydraulic forming equipment offers immense scope for problem-solving in 'A' level work, the problems of designing ECM equipment for school use are equally interesting and stimulating. They include:

(a) Arranging a controllable method of lowering the microbore tube.
(b) Monitoring and keeping the optimum distance between microbore tube and workpiece. (Current consumption rises as the distance decreases so an ammeter in the circuit, for example, can be calibrated to indicate the size of the gap.)
(c) Avoidance of splashing of electrolyte as the hole forms.
(d) Firm holding of the workpiece.

All of the solutions must, of course, take account of the fact that the electrolyte is highly corrosive.

Metal-forming: some chemical techniques

Chemical milling

The history of chemical milling is a long and varied one. Chemical etching techniques were used extensively in the decoration of medieval armour, and have featured prominently ever since in decorative metalworking. It is only more recently that etching techniques have been applied to specific engineering problems, and hence the term *chemical milling*, like 'ECM', has now become a familiar engineering term.

Most CDT specialists will already be familiar with the basic principles of chemical milling, but not aware perhaps of its widespread application in industry for 'machining' large components. In aerospace work, for example, aluminium and titanium are precision milled by stopping out with a suitable maskant and tightly controlled exposure to a suitable etchant.

The advantages of chemical milling (like ECM techniques) over other machining methods are several. It is possible to remove material without generation of heat and mechanical distortions. And this can be done at relatively low cost while ensuring high-dimensional accuracy and a good surface finish. In fact, chemical milling makes it possible to produce components with a large surface area but having narrow and complex cross sections (e.g. aircraft skin panels) which cannot be achieved by other means.

In a typical chemical milling operation, conventional manu-facturing methods are used to produce the overall form of a component which is then reduced in some or all of its dimensions by selective chemical removal. This means, of course, there are no cutting forces, no workpiece clamping problems and few difficulties in accommodating very large and sometimes flexible components. (For all these reasons, chemical milling is used – albeit rather crudely – to reduce the weight of high-performance car engines and even entire car body shells. These are simply immersed in a large tank of etchant until the desired amount of material is shed.)

There are several ways of stopping out a workpiece against chemical action. The most common 'craft' technique is to apply stopping-out varnish, but in industrial operations involving large work, a maskant skin (e.g. neoprene elastomer) is sprayed over the entire surface and then selectively removed by first cutting through in a procedure known as *scribing*. This is often done with a sharp scalpel guided by a template, but it requires considerable skill since excessive pressure on the scalpel will score the metal and give rise to a small step along the fillet edge when etched (fig. 54). After scribing is completed, the unwanted maskant is carefully peeled off and the remaining material repaired where necessary.

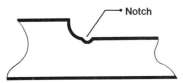

54 *Part-section through an etched metal sheet showing the unwanted notch caused by cutting into the metal surface during scribing. (See fig. 55 for correct fillet geometry.)*

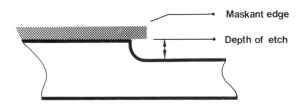

55 *Part-section through an etched metal sheet showing eat-back under maskant edge and showing the rounded fillet profile.*

Under normal circumstances, etching will produce a certain degree of undercutting or *eat-back*. The ratio of eat-back to depth of cut can be established quite accurately for a given set of conditions and vary considerably from the 1:1 ratio seen in fig. 55. By its very nature, the process imposes limitations on the geometry of the edge (or *pocket*) created by etching, but in general the rounded profile shown in fig. 55 is an advantage in components subjected to stress. Many products can be produced in a single etching operation, but if a more complex cross-section is required, several masking and etching stages may be required; this is known as multi-step profiling or etching.

In school, where etching has long been used in the context of decorative metalworking, the concept of chemical milling can be introduced in a range of interesting work which as well as offering opportunities for problem solving, illustrates commercial procedures and production methods. Very thin sheet material such as brass schimstock or cheaper thin-gauge aluminium can be etched through completely to produce small products such as drawing templates (e.g. map outlines for history and geography; symbols for science), or heavier-gauge materials can be selectively *milled* to an appreciable depth for relief detailing (e.g. track patterns for rolling ball-bearing games; surface relief detailing on aluminium pen barrels). The problem of a suitable maskant for school work is solved by using strips of 'Sellotape' which, although not as tenacious as commercial maskants, works surprisingly well as such.

The first stage in producing a simple drawing template using – say – 0.5mm thick aluminium sheet, is to clean the material thoroughly in a detergent solution. When totally dry, a backing sheet of self-adhesive (book) covering material is applied to one entire side of the work and strips of 'Sellotape' to the other. The tape has to be laid down carefully to avoid entrapment of air, and this is best achieved by pressing it down in small increments as it is peeled from the roll.

The scribing can be done either freehand or by using a thin plastic template – which is well worth making up if several identical etched components are required. If possible, exposure of large unstopped areas to the etchant should be avoided to economise on ferric chloride, and where only a small part has to be taken from a much larger sheet, the maskant can be doubled scribed with two parallel cuts so a narrow strip of maskant is removed as opposed to the whole of that covering the unwanted area.

Finally, the workpiece is immersed in a solution of ferric

chloride made up to give a cutting speed of at least 0.25mm/hour. The bath should be agitated if possible by a mechanical stirrer to reduce the problem of gas bubbles adhering around the maskant edge. If this is not available, the work should either be shaken periodically or lightly brushed over in the traditional manner of etching. When the workpiece has been physically separated from the remaining sheet, it is simply peeled off the plastic backing.

Photochemical milling (PCM)

An alternative method of stopping out the workpiece for chemical milling is by the application of a photosensitive maskant or *resist*. This technique, too, has a surprisingly long history, with the French photographic pioneer Niepce producing good photo-etched printing plates as early as 1826. The process has since been applied extensively in manufacturing, and the use of photosensitive etch-resist emulsions is now commonplace in the production of etch-resist images on metals, plastics and semi-conductor substrates.

In essence, photochemical milling involves the application of a photo-sensitive resist to the whole of the workpiece surface, exposure to ultraviolet light (U/V) through a transparent mask bearing the required image (as on a black and white photographic negative) and then 'development' to remove unwanted resist prior to etching. In the case of a *positive* etch-resist emulsion, this remains intact as a tenacious etch-resistant film where U/V is prevented from falling on it by opaque areas on the mask. But U/V passing through the transparent areas causes a reaction which results in the emulsion subsequently dissolving away in the developer. The image on the mask is thus transferred to the work as a precisely corresponding pattern of resist.

The mask is normally produced photographically from a much larger piece of original artwork which can be dramatically reduced in size. The technique is therefore ideal for the production of finely detailed components generally, and essential to the production of minutely detailed integrated circuits.

The technique of PCM can be demonstrated effectively in school providing a safe U/V source is available, or one is prepared to depend on bright sunlight. Commercial U/V light boxes are available for this work, but a low voltage fluorescent unit can be made up at very little cost.[6] (NOTE: U/V radiation can be harmful, and any source must be properly shielded and used with caution.) Photosensitive emulsions can be purchased in liquid or aerosol forms, and some materials can be bought in already photosensitised. Unfortunately, buying the emulsion as a liquid is expensive (i.e. large minimum quantities), and the chemicals involved are toxic. They can also be extremely difficult to apply as an even film – a problem which applies equally to aerosols.

For these reasons, and to ensure successful results, only the use of photosensitised copper-clad boards for printed circuit making will be considered here.[7] This does limit the range of possible work but, in addition to printed circuits, a wide range of decorative forms can be created by partially or fully etching away copper foil and shaping the board itself.

A number of routes exist from the inception of an idea to the finished product, but only two will be dealt with here. Detailed instructions relating to the use of particular photosensitised boards can be obtained from the manufacturers or suppliers, and the following description will be confined to one of general procedure.

The most straightforward technique uses a mask created directly onto the thin acetate sheet used for overhead projector work. This can be suitably stopped out with ink, by the application of rub-down dry transfers such as 'Letraset', or by any equivalent means. When the mask is complete, the emulsion surface of the circuit board is exposed through it to U/V – typically for a period of four minutes. The resist emulsion is then developed (e.g. using a weak caustic soda solution) leaving only an etch-resist image on the copper foil as required. Finally, the exposed foil is etched away in ferric chloride, etched electrolytically in the copper/acid bath (see above), or even plated to build it up in relief.

An alternative technique, and one closer to commercial reality, involves drawing up artwork on a relatively large scale and photographing it to produce a 35mm negative which becomes the mask. If a camera and simple developing facilities are available, 36 such masks can be made in a single operation. The film required to give enhanced black/white contrast is Kodalith 2556 which is very cheap per frame in spite of the initial cost of a large roll. A cheaper alternative is ordinary black and white film exposed and developed for maximum contrast.

Forming plastics

Injection moulding

Injection moulding is a production technique of inestimable importance, and understanding it is the key to understanding the design and performance of an ever increasing variety of injection-moulded products. Unfortunately, many courses in schools, even at 'A' level, can do little more than include a theoretical look at the concept of injection moulding since even small commercial equipment represents a major expense. Attempts to build injection moulding machines have met with varying degrees of success, but given the current constraints of staffing and timetabling, it hardly seems realistic to think that many could now attempt such a project.

It is possible, however, to build a *transfer moulding* device for a fraction of the cost of a commercial injection moulding machine – and within the space of about one hour. Such equipment, used for injection moulding, has been proven in schools and colleges, and many CDT specialists have demonstrated how practical injection-moulding work can become a meaningful part of a CDT programme.

One version of an alternative injection-moulding machine, using the transfer moulding principle, is illustrated in figs 56/57. It consists essentially of a piston/cylinder unit which is charged with plastics material and pre-heated in an oven. When the correct temperature has been reached, the fused contents are discharged under pressure into a mould by forcing down the piston. In this particular example, a mastic gun frame is used to load the piston; the mechanical advantage is considerable, and the whole unit is, of course, portable.

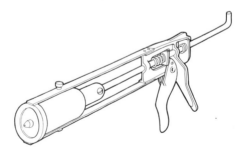

56 *General view of an injection moulding device using a mastic gun frame to eject fused plastic material from the cylinder*

57 Sectional view of the injection moulder showing thickness of cylinder walls in relation to bore

Figs 58/61 illustrate a sequence of operations using this device for moulding a small wheel hub. The cylinder is first detached from its frame, charged with material and placed in a temperature-controlled oven together with the detached piston. When the cylinder contents are fused, the whole device is re-assembled using tongs and gloves. The mould is clamped together and placed on the floor so that the full weight of the operator can be brought to bear on the frame, thus keeping the cylinder nozzle in tight contact with the sprue opening. Immediately this contact is achieved, the mastic gun is rapidly pumped to expel the plastic.

If the cylinder wall is sufficiently thick, enough heat will be retained from the initial heating to fuse an additional one or two charges, but time must be allowed for the charge to 'soak'. This is necessary because, unlike a commercial machine, the plastic only takes its heat by conduction inwards from the cylinder wall, and there is a danger of material at the centre remaining unfused and blocking the injection nozzle.

Low-density polythene is an easy material to injection mould, and if this is not available in a suitable granular form, the powder normally used for plastic dip coating is a good substitute – albeit one that has to be rammed into the cylinder to get a worthwhile amount in. To facilitate easy transfer of the fused plastic from cylinder to mould, both the nozzle orifice and the die sprue are made larger than normal – e.g. 3mm diameter. Although this means there will be a larger blemish when the moulded product is trimmed, it is more than compensated for by the ease of operation.

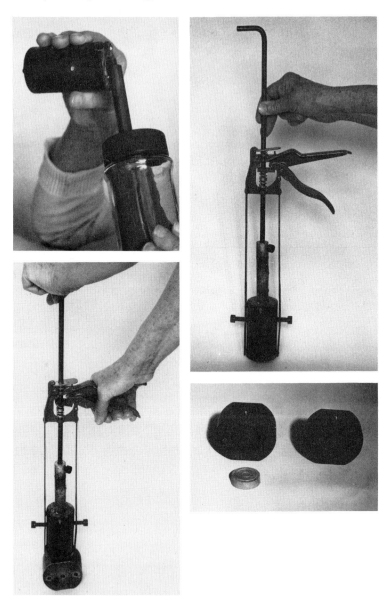

58/61 *A sequence of operations for producing a draughts piece using the injection moulding device illustrated in figs 56/57. The cylinder is charged with polythene dip-coating powder, heated and then re-assembled in the frame for ejection of the contents into a mould*

The overall dimensions are not critical, but those indicated in fig. 57 appear to give optimum results. Decreasing the diameter of the piston increases the pressure obtainable for a given loading, but the size of charge is correspondingly reduced. The piston does not have to be an exact fit into a reamed cylinder because the fused material itself will effect a seal as it rises between the two. But the operator must therefore remember to withdraw the piston while the whole device remains hot.

It should be stressed that any work with fused plastics is potentially dangerous. Under no circumstances should the cylinder be heated unless the piston is withdrawn, and the heating should be done in a temperature-controlled oven at an appropriate temperature for the plastic. Throughout the operation, it is strongly advised that foundry clothing is worn – i.e., gloves, apron and face visor.

Having produced equipment for the injection-moulding operation itself, the question remains of what to do with it. It is often objected that the design and manufacture of dies is beyond the capabilities of most pupils and that therefore even commercial equipment is useful only for the occasional demonstration. There are, however, at least two answers to the apparent difficulty of die making in school.

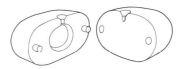

62 *A simple injection moulding mould capable of accommodating cylindrical inserts*

The mould shown in fig. 62 is in two parts and has a plain cylindrical cavity capable of accommodating inserts that can be loosely fitted into one half of the shell. The mould can be made up for class use in a reasonably short time, and once it is available pupils can quite easily design and make inserts for moulding distinctive games pieces, wheels for models, etc. This is a compromise solution, but one which nevertheless involves children in significant and meaningful problem-solving.

On the other hand, it is possible for quite young children to produce an entire metal mould without recourse to any machine tool other than a pillar drill. *Laminated* moulds are quite common in industry, and consist of a sandwich of blanked-out plates that

Manufacturing technology

collectively make up the required cavity. A very simple laminated mould is shown in fig. 63 comprising three plates cut from standard aluminium stock (6mm × 40mm). Aluminium is used widely in commercial prototyping as a 'soft-tooling' material, and it offers many advantages in school work, though similar moulds can of course be manufactured from mild steel.

63 Three-part laminated injection moulding mould in aluminium

All three plates are first cut to a similar size, and the centre plate is drilled and filed out to the required cavity form. One outer plate is then clamped to the centre one so that a sprue can be drilled vertically between the two; but if all three plates are held together at this point, a pair of holes for registration pins can be drilled through. The two side plates may either be left plain or they could be punched, for example, to give relief detailing on the moulded product. If aluminium is used, the plates can of course be chemically milled by methods described above.

Vacuum forming

Commercial equipment for small-scale vacuum forming is available at quite low cost, but it is still the exception rather than the rule to have such a facility in a school CDT department. As in the case of injection moulding, it is not uncommon to find examples of improvised equipment that function well but which take considerable time to build. It is similarly possible, though, to construct an effective vacuum-forming machine at hardly any cost and in a very short time providing that certain limitations on the size and gauge of sheet to be formed are acceptable.

Stripped of any normal commercial refinements – e.g. rising table, plug-assist, etc. – a vacuum former comprises three principal parts: draw-chamber/sealing frame, heater, and vacuum pump (with or without receiver). If these are looked at separately, it is clear that there are several alternative ways of

64

providing each function, and the problem really becomes one of selecting an appropriate set of solutions for a particular requirement.

Fig. 64 shows a complete vacuum former for shallow forming of thin-gauge material (e.g. 0.5mm polystyrene) up to 150mm × 150mm in size. It brings together several unusual but very effective methods of providing the required functions, and versions of it can be constructed quite easily by older pupils to solve specific problems.

The main chamber in this example is made from 12.5mm plywood, and it has a flange and corresponding top frame of the same material. The frame bolts on to the flange, trapping the plastic sheet against a seal of old inner-tube material. This could alternatively be foam rubber draught insulation, but care must be taken in either case to prevent overheating of the sheet and fusing to the seal. Some leakage can be tolerated, but it depends on the means for obtaining a vacuum.

Several possibilities exist for creating a vacuum under the sheet, the most obvious of which is a good vacuum cleaner switched on at the critical moment. For higher vacuums, a majority of commercial machines employ a pump to exhaust a separate receiver which is then opened up to the draw chamber via a valve. The vacuum former illustrated uses this system, but takes advantage of an unusual type of pump which is both cheap and contains no moving parts (fig. 65). This can be obtained in several versions all of which are normally described in supply catalogues as *filter pumps* because they are used to speed up chemical filtering operations.[8] Although the precise design and

64 The draw-chamber and receiver of a small vacuum former

operating principle of these pumps varies, they all depend on a reasonably high pressure supply of water which, passing through the pump, causes air to be drawn in through a third port, entrained in the flow, and expelled with the exhaust water. When such a pump is run from a cold water tap, it is possible to obtain a vacuum of 80/90% if the mains water pressure is good.

In our example, a very efficient filter pump is connected to one end of a receiver consisting of a length of PVC downpipe plugged at both ends. The opposite end is connected to the draw chamber via a valve normally used in pneumatic control. (If such a valve is not available, the flexible PVC or rubber between receiver and draw chamber can be temporarily shut at some point by a single tight fold pinched between two fingers. The 'valve' is opened by suddenly snapping the tubing out straight again.)

In use, the pump should be left running after the receiver is exhausted in order to compensate for any leaks in the system – an essential requirement anyway if the pump is not fitted with a one-way valve. Admission of air to the receiver will create a sudden drop in pressure under the plastic sheet, and for optimum performance the cubic capacity of the receiver should exceed that of the draw chamber by two or three times.

Heating of the plastic sheet can be accomplished by several methods which *avoid* the need for improvised mains equipment. A pre-heated plate brought into close proximity to the clamped sheet is an effective way of softening very thin-gauge material. Ideally, this requires something like a plate of aluminium up to 12mm thick heated either in an oven or in a brazing hearth and then placed over the sheet – resting, for example, on the heads of four panel pins driven into the top frame. These prevent the hot plate coming into contact with the frame and give room for some buckling of the sheet as it heats up and expands. Failures occur, if at all, either because the heating plate is not hot enough or because the frame thickness puts too much distance between plate and plastic sheet – or both.

Alternative means of heating can include the use of an oven with interior radiant heater or approved hot-air blowers normally used for localised softening of plastics. Hot-air paint strippers are being used as substitutes for the latter, but the risks of applying them in this way must be underlined. The nozzle temperatures of electric paint strippers are very high, and few of those available have any sort of thermal overload cutout to prevent serious overheating when the nozzle is restricted.

Working wood: the pneumatic press

The discussion has so far centred on techniques for forming metals and plastics, but there are obviously no limits in applying the 'alternative technology' approach elsewhere. This section concludes, by way of example, with a solution to the problem of providing a suitable press for veneering or laminating in school.

Some departments are fortunate in having access to commercial bag press or mechanical equipment for applying pressure over a large area, but this is rare and attempts are often made to improvise the former with a polythene bag and a vacuum cleaner to exhaust it. Anyone who has tried this will know the problems; at best, the vacuum is low and the vacuum cleaner gets hot. Almost all machines rely on a throughput of air for cooling the motor, and when this is restricted for any length of time – even accepting that air can be bled in – the motor can burn out within minutes.

An alternative way of exhausting the bag is by means of the filter pump discussed above. This gives excellent results, and it is not necessary to seal the bag perfectly providing that the water can be left running throughout the gluing operation. (This is not

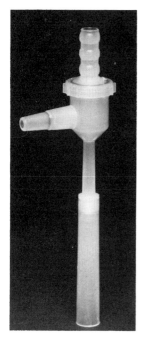

65 *A highly efficient filter pump suitable for exhausting a small vacuum former receiver or a bag for press-work. (See suppliers.)*

*66 Two common plumbing fittings for connecting a polythene bag
to an exhaust pump*

excessively wasteful since the pumps will typically pass very little
water.) The main disadvantage of the system is the problem of
initially exhausting a larger bag since the rate of air removal is
quite slow; the solution is to press the bag as flat as possible
around its contents prior to pumping.

The bag can be constructed using heavy-gauge polythene sheet
sealed on three sides with adhesive tape. When the contents are
in place, the fourth side is closed by folding over several times
and clamping between two stout lengths of wood. The more
difficult problem of effecting a good seal between the bag and its
air exhaust tube is easily solved by using an inexpensive
plumbers' fitting having a wide flange (fig. 66).

Notes

1 See, for example: E.P. DeGarmo, *Materials and Processes in
Manufacturing*, Collier-Macmillan Publishers, 1979. J. Frados (ed.),
*Plastics Engineering Handbook (of the Society of the Plastics Industry,
Inc.)*, Van Nostrand Reinholt Co., 1976, etc. W.T. Harris, *Chemical
Milling: The Technology of Cutting Materials by Etching*, Oxford:
Clarendon Press, 1976.

   The first of these publications is comprehensive, well illustrated, and
includes a good bibliography. The second is more specialised, but
valuable as a definitive general reference handbook on forming
plastics, tool making, etc. *Chemical Milling* is one of a series on
advanced manufacturing techniques, and is an eminently readable in-
depth guide to this important technology.

2 Supplier: Fry's Metals Ltd, Tandem Works, Christchurch Road,
Merton Abbey, London SW19 2PD.

3 Details of 'Prescollan' polyurethane forming rubbers can be obtained
from: The Dunlop Co. Ltd, Polymer Engineering Division, Evington
Valley Road, Leicester, LE5 5LY.

4 As a matter of good practice, no chemical should be regarded as

harmless – or even relatively so. Where a plating bath is involved, all precautions must be taken to ensure its safe use: ventilation should be adequate (or a fume cupboard used), protective clothing – including approved goggles – must be worn, and electrical supplies should be of approved types and safely positioned.

Chemicals can be obtained from: Downswood Products Ltd, Park Lane. Knebworth, Herts, SG3 6PJ. Canning Ltd, Greenhill Crescent, Holywell Estate, Watford. Canning Ltd is the largest supplier of commercial plating equipment, electrolytes, etc., and will supply very useful technical literature on request. (See also *Canning Handbook on Electroplating* – current ed.)

6 Suppliers of low voltage U/V components (and much else): Service Trading Co., 9, Little Newport Street, London WC2H 7JJ. (Tel. 01 437 0576) or: Department PL, 57, Bridgman Road, Chiswick, London W4 5BB. (Tel. 01 995 1560)

7 Suppliers of photosensitised boards: RS Components Ltd, P.O. Box 99, Corby, Northants, NN17 9RS. Maplin Electronic Supplies Ltd, P.O. Box 3, Rayleigh, Essex, SS6 8LR. (Both suppliers sell PCB accessories and a U/V light box unit.)

8 Filter pumps can be obtained from Downswood Products Ltd (see 5 above).

Section 3

Product modelling

The purpose of this section is to examine in some detail a range of product modelling techniques, particularly those which lead to the creation of prototypes whose appearance corresponds closely to the intended commercial product. There are many ways of producing prototype models of this sort, and it would be rash to talk about definitive methods; indeed, professional model-makers must continually invent new techniques to answer particular needs. But there are some methods and materials quite commonly employed, and it is these – together with a number of ideas for getting realism into product models at low cost – that are discussed here. Much of what follows is relevant to three-dimensional modelling in general, but because of its importance as a manufacturing method (and the restrictions on space here) special attention is given to modelling those products intended to be injection moulded.

Not all product modelling is of the exact 'lookalike' variety, but those models intended to convey as faithfully as possible the designer's intentions will typically incorporate precise detailing, exactness of finish, and even realistic weight and 'feel'. The main requirements of this sort of model are therefore *verisimilitude* (the quality of appearing true or real), *durability* – consistent with use as a prototype – and, as appropriate, *performance*. All of these may be crucial in accurately representing an idea for a product when important and often expensive commercial decisions have to be taken at the design stage.[1]

Although we shall be concerned here principally with modelling techniques, it is clear that the design of any product destined to be injection moulded must recognise the constraints as well as the possibilities of the production process itself. Every opportunity should be taken therefore to develop an awareness of what

70

these constraints and possibilities are – something which is almost impossible to do effectively by studying textbooks.[2] It requires, among other things, a serious look at concrete examples, and it is valuable (if not essential) to gather together a range of discarded injection-moulded products for pupils and students to examine and discuss. A good selection of these will confirm, for example, that injection moulding abhors 'undercutting' and favours the use of structural expedients such as webbing to confer strength. It will also show, for example, how a minimum number of moulded parts can be used to make up a complete shell or case as well as illustrating the variety of often ingenious ways employed to fix such components together.

Model-making is largely about creating illusions, and to what extent any product model needs to be structurally similar to the commercial end product depends on its precise purpose as a model. In the case of a torch, for instance, which is not actually required to work at this stage, the model might be solid but given surface detailing suggestive of a two-part moulding. On the other hand, a functioning model would have to contain batteries, etc., and it might well make sense to create a two-part shell corresponding to the two parts of the product as they would be injection moulded. Even so, such a model could ignore internal details unless – perhaps in a separate model – it became necessary to illustrate precise details of battery fastenings and other functional details. There are no hard and fast rules for determining how to proceed in this respect, but particular consideration will be given here to models whose outward appearance is exact and which are constructed to function – especially where this depends on electronics (see *Introduction*).

Materials for modelling

A wide variety of materials can be used in the construction of models which meet the three criteria, but not all of these are readily available for school use. Suitable materials which can be bought in quite cheaply, and are already familiar to most CDT specialists, are aluminium, acrylic and polystyrene.[3] Used individually or collectively to create the form of a model finally sprayed with cellulose paint, these materials provide the necessary foundation for even the most sophisticated modelling. Although materials such as plaster and wood can be used very effectively, they do present additional problems in working – not the least of which is obtaining a suitable surface finish. Some trial work with sprayed acrylic, polystyrene or aluminium will leave few people in any doubts about the speed of working which is possible and the potential for getting highly professional results.

Acrylic

Most CDT specialists will already be familiar with working acrylic sheet, but it may be useful here to rehearse one or two of those techniques particularly relevant to modelling.

The basic structure of a model can be formed from acrylic either by straightforward fabrication (including elements which might be heat-formed), or by machining from a solid block – or both. Whatever method is adopted will depend largely on the nature of the model. Only very exceptionally, for example, would it make sense to waste away massive amounts of material from a single block to create a large rectangular shell. But for very small work – ranging in size between a cigarette lighter and a small calculator – machining from a solid block can be much easier and cheaper than fabrication. (Another factor in school work is the availability of machine tools.)

Acrylic in thick sections is expensive, and it will almost certainly be necessary in school to laminate the sheet material that is available cheaply. In laminating by adhesive bonding, a choice has to be made between different types of 'Tensol' cements – in particular, between one and two-part adhesives. The latter is advantageous since it cures *chemically* at the same rate throughout the bonded area, unlike 'Tensol 12', for example, which ultimately relies on evaporation. (This choice is already made by some LEAs which proscribe the use of two-part cements because of the toxicity of the catalyst.)

Whatever adhesive is used, it is a good idea to lightly key the bond-surfaces of the laminates using fine wet and dry paper (600 grit) and during bonding to apply only pressure consistent with bringing the parts into good contact. Too much pressure may result in damage since the adhesive can facilitate cracking in regions of stress. Full curing must take place before machining through a laminated block, mainly for reasons of safety but partly because the relatively soft lines of material at the laminate interfaces will tend to be dragged out by premature hand working or machining. Any small grooves left in this way can be filled, but the problem is largely avoided by leaving the block to cure for a few days rather than a few hours.

An alternative method of producing the laminated block is by heat welding. This is a less familiar technique than the above, and involves heating the acrylic laminates in an oven to approximately 180° and then clamping them together under considerable pressure. This method has the advantage of being quick and clean, but great caution is required in machining the finished block since it can de-laminate without warning. (Whenever laminated acrylic is machined, it is prudent anyway to clamp the block so that pressure is applied against the laminate faces.)

The laminates may be heated separately and then pressurised between a pair of pre-heated steel plates, or they can be assembled and heated together, pressure being applied when the whole mass is up to temperature (remembering that acrylic is a poor conductor and heat penetration to the centre of a sizeable block will be slow). After pressure is applied – using 'G' cramps, for example – the block can be 'soaked' in the oven for a brief period and then allowed to cool slowly. It should be noted that a considerable loss of thickness occurs in the block as the rubbery mass squeezes outwards under pressure, and it is very easy to end up with a wedge-shaped block if the cramps are tightened unevenly.

The edges of acrylic sheets or small blocks can be prepared for fabrication either by machining (e.g., vertical milling) or by hand methods using simple jigs. One such method is illustrated in fig. 67 which shows an acrylic edge being abraded on wet and dry abrasive paper spread over an old surface plate (or any flat surface – e.g. acrylic). In order to get an edge at 90° to its face, the acrylic is kept in contact with an old angle plate as it is abraded. Wet and dry paper cuts extremely efficiently when flooded with water, and it must be stressed that simple as this technique appears, it is a most effective way of getting high-dimensional accuracy and avoiding the use of machine tools. Fig. 68 shows another popular method of simple jigging when using a file. Here, the acrylic is held in a vice between two mild steel 'jaws' accurately set as guides for the file to work down to. A disc sander can also be used, but it must be remembered that the acrylic will fuse very easily and the dust can be hazardous.

Acrylic sheet can be heat-formed in several ways, and line-bending, for example, might seem an obvious way of making some model parts. This is certainly the case, but it should be pointed out that in exact modelling this is often more trouble than it is worth because of the distortions that can arise; 'kick-up' material has to be trimmed off, for instance, and very accurate handling of the material is called for to obtain a desired radius of bend. Fabricating two pieces of acrylic and then removing material to create the radius can be a far easier option.

73

67 Obtaining an edge at 90° to the face of acrylic by rubbing down on wet and dry paper with an angle plate as a guide

68 A common jigging arrangement for obtaining an edge at 90° to the face of a piece of acrylic: the material is held between two carefully positioned steel cheeks and filed down to them

The acrylic parts can be bonded using one of the 'Tensol' cements – with the same jig arrangement shown in fig. 67 employed to ensure accurate edge jointing. Mechanical fastening is also a possibility since acrylic can be tapped and screwed into; this is an especially useful option where dissimilar materials have to be joined to the acrylic.

Polystyrene

Polystyrene sheet, familiar in CDT work as a vacuum-forming material, is the staple diet of many small-scale modellers and is an extraordinarily versatile material in the context of precision model-building. It lends itself to fabrication rather than hand or machine wasting from a larger block, and is therefore ideal for use in school when machinery is at a premium. Polystyrene sheet can be obtained easily in thicknesses ranging from 0.25mm to 3mm. It can be cut using practically any kind of saw, but it happens to be far easier to *score and fracture* than acrylic sheet – and this is by far the best method. Ideally, the sheet is scored using a tool that tears out a string of material to leave a small 'V' groove. A hook-profiled tool will do this, and one can be ground quite easily from a discarded hacksaw blade (fig. 69). After scoring, the sheet is positioned to project over the edge of a table or bench, and cracked. If sheet metal-folding equipment is available, it is an ideal alternative.

69 The profile of a suitable tool for scoring polystyrene prior to fracturing

Polystyrene can be worked in many ways similar to acrylic, but it is far more heat sensitive and when it comes to machining operations such as milling, the result is likely to be localised fusing unless great care is exercised. For this reason, it is advisable to stick to hand methods of working such as the sample jig technique illustrated (fig. 67).

Polystyrene is the material of most plastics model kits (e.g., 'Airfix'), and it can be bonded using one of the many brands of either cement or liquid solvent available for general modelling work.[4] The cement is a solvent type and finally cures by evaporation after fusing with the surface to which it is applied. A quite rigid butt joint can be effected in minutes because of the speed of curing, and this can be a great advantage in speeding up assembly time of a model. However, the cement must be applied quickly to avoid formation of a superficial skin. As in the case of acrylic, *total* curing by evaporation is a slow business, and the joint should be left as long as possible before further working to avoid preferential removal of soft material along the join lines.

Because it evaporates so rapidly, liquid polystyrene solvent is rather difficult to use for bonding of edges, but it does have the big advantage of not leaving behind a fillet of excess cement squeezed out when parts are brought together. The solvent can be used in a number of ways – probably the most common of which is application with a brush to each of the two surfaces to be joined. (It should be noted that fumes from either the cement or the solvent can soften the material, and if an enclosed box is put together without ventilation, it may distort.)

Constructional techniques

When a vertical milling machine or equivalent facility is available, it is very straightforward to produce a small model – or some detailed component of a larger one – simply by cutting away from a solid block of acrylic or aluminium. If a small box or case is required, for example, a standard length or long-reach slot drill can be employed to mill out the cavity and a standard end mill cutter to face off exterior surfaces. Where acrylic is used, though, the walls should be kept as thick as possible for ease of machining and strength.

Most modelling, however, will involve an element of fabrication, and fortunately the two plastics considered here can each be bonded to give the equivalent, more or less, of a weld. Very durable models can therefore be constructed using nothing more than butt joints with web-type reinforcements added where appropriate. The one cardinal rule here is that the surface area at each join interface is sufficient to confer reasonable strength, and this normally means using at least 2.5mm gauge sheet if polystyrene is used.

Additional strength can be provided by webs, metal pins or screws (as wall thickness permits), polyester resin filler paste or even polyurethane foam – LEA regulations permitting. Products such as 'Isopon', normally used for car body repairs, are commonly used in this type of modelling for both structural work and surface detailing. Filler paste, for example, can be run around the inside corners of a box to provide thick fillet reinforcement to the butt joints, and this might at the same time be used for locking in interior components. One-part polyurethane foam is now commonly available as a DIY material in a pressurised can dispenser, and in this convenient form it can be injected into a box to fill any unused interstices. This confers considerable strength without paying the penalty of too much additional weight. In fact, the entire cavity of a box or case model can be filled with the foam which is then carved out to provide the necessary accommodation for components and fittings – e.g., batteries. Where its use is permitted, all necessary precautions must be taken in the use of polyurethane foam, including protection from dust when it is abraded.

Vacuum forming is an important technique for the model-maker, especially for larger-scale work; it is not, however, the panacea that many of those new to exact modelling seem to imagine. It is invaluable, for example, when there are several identical components to produce or when a thin curvilinear shell is required. But it has to be remembered that a former must be constructed in the first place and that it has to match the standard of accuracy required in the complete model. This takes up time, and quite often that can be better invested directly in constructing the part by fabrication or wasting from solid.

In dealing with products involving electrical or electronics packaging, the constructor concerned with *working* models is faced with some interesting problems not encountered when the model is largely visual or static. It is not a good idea to seal circuit boards, batteries and other components into the case; some access should clearly be provided, especially for battery replacement. It has already been pointed out, however, that such provision does not necessarily mean building into the model a realistically opening battery compartment. The outline only of the intended compartment opening can be represented with surface relief detailing, and actual access to the battery provided through taking the whole model apart.

There can be no absolute guide about where and how a model should open up, but there are several distinct advantages in dividing or breaking it so that the two or more parts correspond in the way they break to the intended commercial product. If, for example, the intention is to mould components in different coloured plastics, this simply entails spraying the model parts different colours as appropriate. (Pupils and students will probably find two-part models a sufficient challenge to begin with, and it may be worth stipulating in the design brief that the final product is to be moulded in two parts only.)

Many injection-moulded products incorporate a deliberate 'shadowline' at the interface between mating components, often in the form of a shoulder running around one of the shell parts. This may be included to disguise the fact that production tolerances are not set sufficiently high to obtain absolutely accurate mating, or because the result is visually more acceptable – or both. This kind of detailing can, of course, be incorporated into a fairly straightforward model, and in addition to enhancing realism, it also helps to disguise the possibility that tolerances on the actual model will be less than perfect!

A shadowline around the break does not, of course, eliminate the need for accurate registration of parts in the first place. There are several methods for ensuring accurate correspondence which CDT specialists will be familiar with in principle. For example, a small two-part case machined from the solid can be worked in separate parts to produce the interior cavity and then temporarily assembled for exterior machining. If fabrication is called for, another familiar procedure is to build the complete case and then split it – facing off sawn edges, for example, on wet and dry paper.

Where a model is constructed in two or more parts, there must be an effective means of holding these together which may or may not correspond to the proposed commercial solution. Again, it is impossible to generalise on this point, but some common methods can be listed as follows:

1 *Screw fastening* This method is used extensively in commercial products and can be easily applied to models using either self-tapping screws or conventional screws in tapped holes. In either case, great care must be taken working in plastics to prevent bursting out when screws run near a surface. To accommodate them, it may be necessary to leave 'lands' when machining from the solid, or to build up blocks when fabricating, and this requires planning from the outset. Thought given to location of the screws and their possible concealment is of the utmost importance since nothing can be worse than indiscriminately placed fastenings. A close look at some commercial products will show just how cautious designers tend to be in this respect.

   For the purposes of a model, the parts might well be screwed together even though a system of integrally moulded clips might be envisaged for the commercial product. In this case, deeply counter-bored screw holes on the bottom of the model would show up as a necessary evil but least obtrusively.

2 *Pin and socket fastening* Two or more parts of a model can be held together with small-diameter pins set rigidly into one component and locating in corresponding socket holes in another. Suitable pin material might be welding rod, crochet or knitting needles, but because of the difficulty of bonding these in with any degree of permanence, a good alternative is the use of small-diameter set screws screwed into one of the mating parts so that a surplus length protrudes as the pin.

   Perhaps the easiest way for the beginner to ensure accurate

registration of pin and socket holes is to drill through when both components are accurately clamped together. For example, if set screws are to be used as pins, tapping size holes are first drilled through the two clamped parts, but not right through so that one half of the model is left with a series of blind holes to act as the sockets. When the two parts are unclamped, these blind holes are enlarged to accommodate the full diameter of the screws used. The through holes in the other half are tapped and counterbored slightly on the outside surface; the screws are then tightly inserted from this side and their protruding ends (now the pins) trimmed off to length. All that now remains to be done is to fill in the counterbored hole over the set screw head using filler paste.

It might be thought that although the pin/socket system will provide registration, it will not keep the parts of the model together. In practice it will be found that if only one pin of several is slightly out of alignment (usually a natural occurrence) this will give an adequate locking effect. (It should be noted that where a similar method is used to locate smooth pins, these should be bonded into the blind holes so that if they work loose they push against solid material and not filler paste.)

3 *'Velcro' fastening* There are occasions when standard 'Velcro' tape can be used to fasten component parts together. This normally requires a reasonable surface area to attach the tapes and accurate distancing between the tape halves so that the model, when assembled, is not separated by the combined thicknesses of the tapes. This involves recessing them as far as possible, and sometimes relying on a certain amount of 'spring' in the model to get the two tapes to knit together effectively.

4 *Magnetic fastening* Small magnets set into one half of a model and locking onto pieces of steel set into the other may appear an attractive proposition, but in practice a number of problems arise. The magnets must be quite powerful, and these are usually expensive. In addition, it is usually necessary to incorporate some form of positive location to prevent the parts sliding over one another. Magnets can be used to good effect, however, for attaching smaller components such as thin covers, which can be backed with tinplate for attraction.

5 *Adhesive tape fastening* If a model is not to be taken apart too often, double-sided adhesive tape offers an easy solution to the problem of temporary fastening. But, as those familiar with this material will appreciate, it is an excellent adhesive in

its own right and coverage over too large an area will produce a permanent rather than a temporary join. (Double-sided tapes are used for permanent bonding in many commercial products, and it is a most useful permanent adhesive to use in this kind of modelling.)

Surface detailing

In producing the illusion of realism in a model, the importance of accurate surface detailing cannot be underestimated. It is here that the ingenuity of the model-maker is fully stretched in creating surface textures, details such as logos and lettering, and countless other relief effects. For the sake of convenience here, surface detailing techniques will be considered under two headings: *structural* detailing and *applied* detailing. The distinction is a somewhat novel one, and deserves a little explanation.

By structural detailing is meant the manipulation by hand or machine working of the basic structure of the model to produce detailed relief features – for example, material removed from an edge to produce a radius, or a pattern of holes drilled through a surface to form a sound outlet. On the other hand, applied detailing will involve additions to the basic model structure. This could take the form, for instance, of commercial relief lettering bonded to the surface, or a complex build-up of sheet material to create a low-relief image. But it must be stressed that the construction of a single model will typically involve both categories of work, and it is therefore important that any initial structural decisions take into account the kind of detail to appear on the completed model.

Structural detailing

Radius work

External or internal radii (or other curvatures) can normally be taken care of easily if part or all of the model is vacuum formed; necessary details are built into the former using familiar wasting and build-up techniques.

Where the model is machined from solid, or fabricated in thick sections (i.e., anticipating a particular depth of radius), material can be removed either by machine or hand. An internal radius is best accomplished on the vertical milling machine using a ball-nose cutter. Fig. 70 shows a (solid) model having an internal radius running across the centre produced in a single pass using a 12.5mm diameter cutter. A similar radius can be produced by hand working with circular section emery-sticks (covered in wet and dry paper) abrading against a 90° step pre-machined on the model. The larger the radius, though, the more difficult this becomes; unless some kind of guiding jig is devised for the job, the tendency is to rock the stick and enlarge the radius at either end.

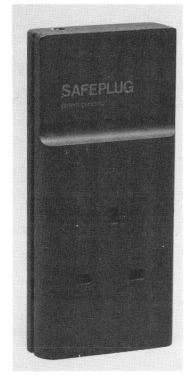

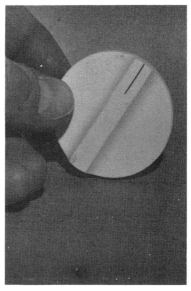

70 (left) A solid model machined from laminated acrylic

71 (above) Control knob machined from acrylic using a ball-nose milling cutter

Unfortunately, a range of milling cutters capable of producing an equivalent external radius is not commonly available. One substitute is a router cutter held in a Jacobs chuck, providing that the chuck arbor is held in the machine with a drawbar to prevent it coming loose through vibration. (Any such arrangement would naturally require stringent attention to table guards, etc.) Unlike internal radiusing, however, it is quite easy to obtain an accurate external radius by hand. A good method is to sweep the entire edge of the model across a sheet of wet and dry paper. It is quite straightforward, of course, to produce a small radius or chamfer using emery sticks, files – or both.

Reliefs and textures

Although available milling cutters and the thickness of material used in the model set definite limits, the vertical milling machine can be used to produce various kinds of reliefs and textures. Even though it may sometimes appear wasteful, practically any size detail can be projected into relief on a surface simply by removing all the

72 *A sampler illustrating how matrix board can be incorporated into a model as a speaker grill*

73 *An acrylic pressing for a lens cover and the impression tool used to produce it. The tool consists of a series of welding rod stubs bonded to a steel plate with 'Araldite'*

material around it. Fig. 71 shows an example of a control knob machined in this way using a ball nose cutter to create the radius.

Coarse textures can be produced by making a series of parallel cuts with a small end mill, leaving a regular width of material between each. If the work is turned through an angle after the first series of cuts and the operation repeated, the kind of square or diamond hatch texturing frequently seen on camera cases and photographic equipment can be produced quite effectively. Similar effects are possible using only the tip of a drill to create a regular pattern of dimples, but precise depthing is essential to this procedure. If it is used, the tedium of accurate indexing over the surface may be avoided by using matrix circuit[5] board temporarily fastened over the workpiece to act as a drill guide or template. (These boards are thin paxoline sheets pierced with a matrix of regularly spaced holes to accommodate pins for mounting circuit components.)

Matrix boards are exceedingly useful too as drill guides when a pattern of holes has to be drilled through a surface to provide a sound outlet. Fig. 81 shows the result of this technique in producing a multiple hole grid on the front of a timer. To avoid drilling very large areas, the matrix board itself can sometimes be incorporated (fig. 72).

The heat-memory potential of acrylic sheet can sometimes be exploited to produce relief detailing, and fig. 73 shows a translucent lens cover for a lamp in the course of manufacture using this familiar technique. However, it must be remembered that an impression tool has to be made up in the first place and that dimensional accuracy is difficult to achieve on account of unpredictable distortions. For the one-off part, it is usually far easier to machine or fabricate directly.

Applied detailing

A remarkably wide range of materials can be applied to the surface of a model to create various relief details. And when the model is sprayed, these materials will blend together to give the required homogeneous moulded-in-one appearance. Three 'add-on' materials will be considered here: paper, plastics and sheet metal.

Paper

Paper might seem an unlikely candidate for this sort of

modelling, but it can be used to excellent effect in low-relief work. Perhaps the most useful material is the self-adhesive labelling supplied in various shapes and sizes on a wax paper backing. Most stationers sell address labels on rolls of wax paper in addition to a variety of smaller labels and die-cut stickers – some of which, such as arrows, can often be used on models as found. These labels stick effectively to a smooth plastic or metal surface, and the apparently vulnerable edges are in fact sealed down and hardened by paint when the surface is sprayed. Indeed, the whole of the paper surface is surprisingly tough, and particular care has to be taken only when rubbing down layers of paint with wet and dry paper, which must here be used dry.

The paper for application should first be laminated to the required thickness by sticking the labels one on top of another. Although a single thickness of label paper will normally stand clear in relief after spraying, it is usual to apply at least two thicknesses. The complete paper lamination is carefully cut to shape on a smooth surface such as acrylic using a sharp scalpel or knife. This must be very sharp since a good quality cut edge is essential in obtaining sharp detailing on the model. Fig. 83 illustrates the usefulness of built-up paper in creating ribbing on the surface of a model representing a tile cutter. (Note: the paper in this example is laminated to the exceptional thickness of 0.5mm.) Apart from details of this sort, the technique has great potential for the creation of product logos and larger lettering.

Fine surface texturing is a common feature of injection-moulded products and is used – if for no other reason – to impart visual interest to an otherwise plain area. Textures on a model can be produced simply by applying textured papers. A useful material in this context, and possibly the one closest to hand, is wet and dry abrasive paper. It is most effective as a general texture medium providing that only the coarser grades are used; if they are too fine, paint will tend to fill up the interstices and nullify the effect.

Wet and dry paper can be applied using something like spray-mount adhesive either as an area standing in relief or as an insert. The last of these methods is visually more acceptable on the model, and is likely to correspond more closely to commercial practice. If the model base material is acrylic or aluminium, a shallow depression can easily be milled out over a large (flat) area to accommodate the thickness of paper; if the parent material is polystyrene, thin strips of that material can be used to build up an edge surround for the paper insert. But if the model surface is curved in one or more planes, texturing with paper becomes difficult if not impossible. Under these circumstances, grit can be sprinkled on during spraying, but this is a procedure demanding great skill and care.

These considerations also apply to the use of embossed papers obtainable in various patterns and textures for other kinds of modelling work. Applied with spray mount, these can make excellent surfaces providing that the model is not to be handled too roughly. It is possible to make up embossed paper by passing self-adhesive labels, for example, through a rolling mill in contact with a hard textured plate. Figs 74/75 show how an attractive dimpled texture can be embossed on to a paper label by rolling it sandwiched between a piece of the matrix board referred to above and a resilient pad of several layers of scrap paper. (Other textures can be created by substituting for the matrix board prepared plates of etched brass, copper or aluminium. These can be suitably textured by stopping out against an etchant with dry transfers such as Letraset's 'Letratone' range.)

The use of paper gives a lot of scope for ingenuity in modelling, and even what is normally regarded as scrap has many uses. Fig. 76, for example, shows a coarse texture produced from waste material; instead of peeling off the individual die-cut labelling disks from a self-adhesive sheet, these are discarded and the surrounding grid removed for application to the model surface.

74 Rolling self-adhesive paper labels against matrix board to produce a dimpled texture

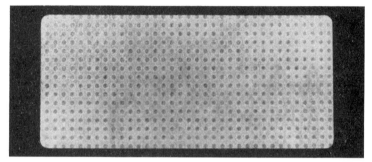

75 *Texture on a paper label produced by rolling against matrix board*

76 *(left) A sampler showing how an interesting texture can be produced by the application of paper*

77 *(centre) Embossed polystyrene sheet from Slater's 'Plastikard' range. (See suppliers.)*

78 *(right) Three-dimensional alphabet produced by Slater's 'Plastikard'. (See suppliers.)*

Plastics

Thin sheet polystyrene (and other plastics) can be similarly used to create relief effects. It can be bonded with spray mount if the area is appreciable, or by using polystyrene solvent cement. Modelling supply firms such as *Slater's Plastikard* supply polystyrene in a number of forms specifically for scale model railway work, and these can be very useful to the product design modeller.[6] 'Micro-line' – thin polystyrene sheet cut into strips of varying widths – is one example. The same firm also supplies a wide range of embossed polystyrene sheet designed to represent surfaces such as cobble stones and roof tiling at various reduced scales. These make excellent surface textures as the sampler shown in fig. 77 illustrates.

Small-scale relief lettering appears somewhere on most injection-moulded products to give instructions, patent identification, country of origins, etc., and such lettering on models – sometimes more than any other single detail – can greatly enhance realism. There are several easy and inexpensive ways of producing low-relief lettering, and one favourite involves the use of Slater's 'Plastikard' moulded alphabets, supplied in the form of complete injection-moulded sheets (fig. 78). These are precision moulded in polystyrene and range in size from 1.5mm to 6mm high.

Precisely how the letters are applied depends first of all on the base material of the model. If this is acrylic or aluminium, the best method is to flood the immediate surface with polystyrene solvent and then lay the letters down using fine tweezers or a scalpel used to spike them. The solvent does not affect the acrylic or aluminium but attacks the back of each polystyrene letter making it, in effect, self-adhesive. Although the solvent rapidly disappears, the letters can be slipped around to their final position before it completely evaporates. The beauty of this technique – which can also be used in the application of other small details – is that it leaves no messy squeezed-out adhesive between the surface and the letters.

The same method cannot, of course, be adopted where polystyrene is the base material of the model; in this case, the liquid polystyrene should be applied to the back of each letter using a brush or by dabbing with a solvent-moistened pad. If a large number of letters are involved, it is advantageous to compose them in reverse and face down on the sticky side of a length of 'Sellotape' and then to offer the entire assembly to the model after application of solvent. In this way, precise positioning and alignment of all the characters is accomplished before fastening down, and the assembly can be offered to the model 'dry' to assess the effect.

An alternative method of lettering is to use 'Dymo' embossed plastic tape, which has the added advantage of being self-adhesive. There is a wide range of inexpensive 'Dymotape' machines to choose from and various sizes of tape. Because the letters appear in relief on tape which itself has an appreciable thickness, the use of this material has to be carefully planned and considered. If simply applied to a surface and sprayed, it still looks like 'Dymo' tape, but if slightly recessed into a shallow trench it finally appears to be part of a complete moulding.

Metals

Perhaps the most useful sheet metal for producing low-relief effects is brass schimstock which can be bonded to practically any surface using spray-mount adhesive. Conventional shearing will, of course, produce edge distortion, and so the best way of 'cutting' is by stopping out the desired shape and etching right through with ferric chloride (see also *Chemical milling*, p. 55). An appropriate thickness of schimstock is selected – remembering that heavier-gauge material will end up with a more pronounced undercut edge – and the stopping-out medium applied. The brass is then securely backed on to a sheet of self-adhesive plastic of the sort used for covering books – or, if the piece is small enough, 'Sellotape'. The purpose of this backing is twofold: it acts as a stop-out on the back of the brass sheet, and it provides a convenient means of handling the work in the ferric chloride bath, an important consideration when it is very small. After all excess brass is etched away, the remaining material is carefully peeled off the backing sheet and attached to the model.

It may be objected that this method of low-relief detailing is a rather cumbersome way of doing what can more easily be achieved with paper. There is some truth in this when the relief is relatively large and simple in outline, but the metal-etching technique is difficult to match for precision especially when the detail is both small and complex. This method also enjoys a unique advantage. Dry-transfer rub-down lettering (e.g., 'Letraset' or 'Edding transfer') happens to be an excellent stop-out against ferric chloride if it is fresh and applied with care to a clean metal surface. It is therefore possible to use it on schimstock to create relief letters (or other graphics) in any style and size produced in a particular transfer range.

False detailing

Before looking at finishing procedures, it is worth underlining here that even where a model is designed to function in some way, not all parts need necessarily work realistically. This point has already been made about battery compartment covers, but the principle applies equally to all sorts of mechanical and electronic control functions. A battery compartment cover might be defined by a narrow milled line, and completed by a couple of applied paper details (see above). The control knob shown in fig. 79 can be produced either by defining the outline with a circular cut (e.g., using a rotary table on a milling machine) or by trepanning out a hole and inserting an acrylic disc.

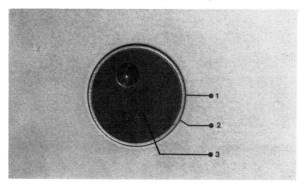

79 *False control knob detail produced with a circular milled cut*

Finishing the model

What distinguishes the exact model above all else is attention to detail and overall quality of finish. The first of these costs nothing except time, and the second can be accomplished without dependency on expensive spray plant. In fact, the use of conventional spraying equipment will not be dealt with here. It can be set up quite cheaply, but to satisfy both LEA safety regulations and the varied demands of model making in schools, investment in specialised equipment can cause more problems than it is worth. Attention is given instead to the cheap aerosol paints normally used for DIY car repairs (*cellulose paint not enamel*). This in no way implies that it is safe to use such materials indiscriminately; spraying of cellulose paint by any method is a potential fire/explosion hazard, and exposes people to a toxic material. It is always essential, therefore, that ventilation and/or extraction is adequate even when using aerosols.

Of the normal aerosol range, matt black is probably the kindest to the inexperienced user – followed closely by the metallic paints. Excellent results can be obtained with the latter, and it is worth noting that many injection-moulded products are given their metallic finish by spraying.

When the model is structurally complete, it should be checked for blemishes such as deeper scratches which are likely to show through the painted surface. As a general rule, it is safe to rub down the whole model with 600 grit wet and dry paper, and this should certainly be used to get rid of marks from previous abrasives. It is difficult, though, to identify all the problems prior to spraying, and the first coat of primer may therefore be regarded as *diagnostic*; any faults highlighted can then be either filled or rubbed down.

It is essential that the model is clean, dry and free from dust prior to spraying (avoiding the use of proscribed de-greasing agents). A wire tether or other jigging method is then devised to support the workpiece so that it can be rotated as the spray jet is swept over the surface. Many children and students view aerosols as an instant answer to good finishing, and they really need explicit instructions in their use. The spray jet must not be initiated or stopped directly over the work; it should be started to one side, swept across back and forth, and finally stopped pointing away to prevent any spluttering on the surface. In aerosol form, cellulose paint is very thin, and must be built up in many thin layers. This is especially important if spraying over polystyrene since the paint will not adhere well unless given the lightest of 'dust' coats first.

After the model has been given its initial coat of primer, it should be inspected for faults. Any small scratches or blemishes can be taken out either by careful rubbing with 600 grit abrasive or filled in using *cellulose putty* wiped across the surface – and then rubbed down. (Fillers such as 'Isopon' are too coarse for this particular job.) More primer is then applied, and the process repeated until the surface is as intended.

When the final top coat of paint is applied, the distance from the jet nozzle to the surface determines the type of finish irrespective of the type of paint. A relatively generous film of paint applied at close quarters will tend to smooth out, but if it is sprayed lightly from a greater distance, the effect is to produce a fine granular texture. It is often desirable to give models representing an injection moulded product either a matt and/or finely textured surface, and so the top coat might simply be left unburnished or uncut. But if consistency of finish is important, consistency of style in spraying is essential.

If two contrasting colours are involved in the finish, attention must be given to masking-out procedures quite early on. Masking may not be necessary at all if the different coloured elements of the model are physically separable, but where this is not possible the whole thing must be sprayed with the dominant colour and masked off for further spraying. The most obvious masking material is tape, but ordinary paper tape is useless for high definition work, and scale-modellers use a smooth plastic tape. Alternatively, plastic masking film can be used – or even a liquid latex. The last of these is a material rather like 'Copydex' adhesive which is applied with a brush and quickly evaporates to leave a latex skin easily peeled off after spraying.

Unfortunately, all of these masking systems will result in a built-up edge of paint partly because of the additional thickness of paint and partly because the paint forms a fillet against the maskant edge. This problem can be mitigated by using a minimum of paint and carefully controlling the direction of spray. It should also be stressed that burnishing the edge of masking tape or film will do much to prevent intrusion of paint and complete ruination at the last minute.

If a 'soft' diffused edge is required between two colours, spraying should be done through a mask of paper or card suspended a small distance from the surface. In this way some paint penetrates under the edge of the mask to soften the contrast. The effect depends largely on the distance between mask and model, but judging this distance and directing the spray evenly requires some practice.

Where very small areas have to be picked out in a contrasting colour, it may be possible to use spirit-based inks applied using one of the new 'airbrush' systems which directs a jet of air over the felt tip of a pen to produce an aerosol dispersion. This technique is illustrated in fig. 80 using a garden spray unit for the air supply.

After final spraying, the model might be finished using rub-down letters, etc. These adhere well to the painted surface providing they are fresh, and give the sharp definition effect of screen printing which is a common method of getting graphics onto commercial products. If used very carefully, felt-tip pens, especially metallic types, can be used to highlight relief details. This can also be achieved with hot-blocking foils which are widely used in commercial products. The foil consists of a thin metal film vacuum-deposited on to a refractory plastic sheet and overlaid with a thermosetting adhesive. The foil is placed adhesive-side down over the work and a hot iron rubbed on the back – activating the adhesive and causing the metallic film to adhere to the surface. (Bookbinders now use this material for lettering, etc., on book spines.)

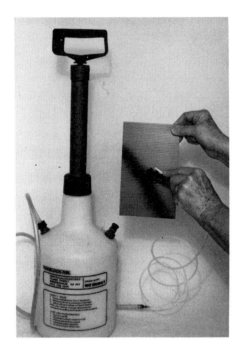

80 An 'airbrush' system using spirit based inks – sometimes useful for small detail work on models

Case studies

Most of the techniques discussed above have been combined in various ways to produce the prototype models illustrated in figs. 81/91. All of these represent, as far as possible, the appearance of the intended injection moulded product.

Timer (fig. 81). This model represents a simple kitchen timer, and is constructed in polystyrene over a flat acrylic base forming the back. The timing circuit is described in *section 4* together with an improvised form of tilt switch used for activating and stopping the timer.

Telephone (fig. 82). Part of a simple intercom system, this model incorporates ex-GPO transducers in the handset. These are energised by a battery contained in the wall-mounting unit which also has two station selection switches. The handset is machined from a solid (laminated) block of acrylic, and the wall unit is fabricated from 6mm acrylic sheet. Mouthpiece and earpiece grills are acrylic milled with parallel slots.

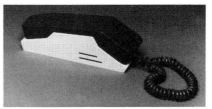

81 (left) Timer model
82 a&b Telephone (intercom) model

Tile cutter (fig. 83). The tile cutter prototype is constructed mainly in acrylic, but functions mechanically as the commercial product would. The tile is pressed against an adjustable straightedge on the board and passed glaze down over a sharp carborundum point which scores a line. It is then turned over and fractured on the metal insert adjacent to the scoring pin. The surface ribbing consists of paper strips built up in layers, and the relief lettering is Slater's 'Plastikard' alphabet. Rub-down transfers are used to give the product a name.

Frame clamp (fig. 84). This is another mechanical model illustrating an idea for a frame clamp for the DIY market. Accommodation for a picture frame is coarsely adjusted by altering the position of corner blocks along the four arms of the clamp, and the pressure is then applied by turning a central boss which pulls all the arms inwards and effects a geometrical lock. This model is constructed entirely in acrylic with the exception of steel pins, etc.

Torch model (fig. 85). The principal part of a multi-function product is represented in the model here. This is the battery unit of a device which can function as a torch when a bulb module is plugged into the end – or which, for example, can be used in testing fuses when a different module is inserted. The model is

93

Product modelling

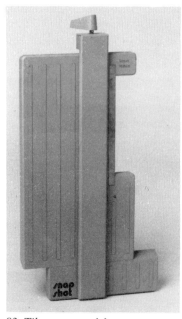

83 Tile cutter model

84 Frame clamp model

86 Buzzer model

87 Hotel door alarm model

85 Torch model

88 General purpose domestic alarm
(model of control box)

89 Domestic alarm system
(control box model)

constructed throughout in acrylic and uses a pin and socket system to keep the three parts assembled.

Buzzer unit (fig. 86). This buzzer unit is one outcome from an introductory modelling exercise involving the packaging of a solid state buzzer for signalling. Construction is similar to the timer (above) and the sound outlet is a matrix board insert.

Alarm systems (figs 87/90). All the devices illustrated here are small self-contained alarms which involve the use of the electronic switch circuit discussed in *section 4*. Fig. 87 shows a combined polystyrene and acrylic model design as a hotel door alarm. A tremblcr switch scts thc alarm off, and it is re-set with a magnetic key. (Both switches were made up in the way described

90 Television viewing alarm: model of self-contained unit

in *section 4.*) Figs 88/89 are both general-purpose portable alarms modelled in polystyrene. Any type of switch can be used to trigger these including, of course, membrane panel types. Fig. 90 is a committed alarm which sounds when a television is turned on. An LDR (light dependent resistor) fastened to the screen takes the place of a triggering switch, and the alarm is turned off by a magnetic key.

91 Audio-reward money box model

Audible moneybox (fig. 91). A view is shown of a polystyrene/acrylic model representing a moneybox which produces an audible 'reward' for any coins inserted. Circuitry and switching is similar to those discussed in *section 1.*

Notes

1 For general information see: B. Hawkes and R. Abinett, *The Engineering Design Process*, Pitman, 1984. E. Tjalve, *A Short Course in Industrial Design*, Newnes-Butterworths, 1979. There is very little literature on the kind of three-dimensional modelling discussed here. Some useful material can be found in model engineering and scale modelling publications – and ideas, as well as useful materials, can be gleaned from good model suppliers. An excellent book, of considerable *indirect* value, is: G. Wingrove, *The Complete Car Modeller*, New Cavendish Books, 1981.
2 For a definitive general reference handbook see: J. Frados (ed.), *Plastics Engineering Handbook (of the Society of the Plastics Industry, Inc.)*, Van Nostrand Reinholt, 1976, etc.

3 Suppliers of polystyrene include: Amari Plastics PLC, 2, Cumberland Avenue, Park Royal, London NW10 7RL. Seawhite of Brighton Ltd, 61, Waterloo Street, Hove, Sussex, BN3 1AA. E.M.A. Model Supplies Ltd, 58–60, The Centre, Feltham, Middlesex.

The last of these firms stocks embossed polystyrene sheet material as part of its remarkable range of plastics materials and mouldings for professional modelling of petrochemical plant, etc. It is a prime source of material for exact modelling.

4 Supplier of multiple quantities: Ripmax Models, Ripmax Corner, Green Street, Enfield EN3 7SJ (Tel. 01 804 8272).

5 Supplier: RS Components Ltd, P.O. Box 99, Corby, Northants, NN17 9RS.

6 Slater's Plasticard Ltd, Royal Bank Buildings, Temple Road, Matlock Bath, Matlock, Derbyshire, DE4 3PG (Tel. 0629 3993).

This firm supplies a wide range of modelling materials in addition to the items mentioned. The service is excellent, and it is well worth obtaining a catalogue.

Section 4

Microelectronics for product design

This section is based entirely on a series of Teacher's Notes written to support an initiative encouraging the use of microelectronic (digital) devices for product design work in the upper school. A range of likely needs was carefully considered, and the circuit ideas presented here are an attempt to satisfy some of the more common ones. All the circuits centre around different applications of a single logic principle, and in practice each can be built using an identical integrated circuit (IC). This was a quite deliberate policy, and perhaps deserves a little explanation since it might be objected that better results can be obtained using more specialised ICs or by modifying and adding to the circuits suggested.

It is true that the circuits – stripped as they are of any refinements – do have some drawbacks, but they seem to be perfectly adequate for their intended role – and they can claim three major advantages: each circuit is *easy to understand* once the basic logic principle is grasped; they give adequate results with a *minimum component count*, and they use a very cheap IC.

Beginniners in electronics, even when slavishly following a given diagram, invariably make mistakes, and many become totally dejected by lack of initial success. Each circuit here is very easy to construct, and experience clearly shows that the improved chances of immediate success greatly enhance motivation.

It would be a serious mistake, however, simply to offer a package of ready-made circuits. Although the electronics might not be regarded as paramount in some product design exercises, there should obviously be some conceptual purchase on underlying principles – ideally to the point where a student or pupil can be said to be working from *first principles*. All the circuits presented here have one logic 'ingredient' in common, and once

98

this is understood, many variations on basic circuit themes can be explored.

Finally, the question of cost. Components such as resistors, LEDs and capacitors remain quite cheap, but the recent crisis in IC supply has pushed up the cost (and availability) of some ICs virtually out of reach for normal school work. Future supply is far from certain, but some ICs remain cheaper and far more available than others – including IC 4001 used in what follows. (And including other ICs that can easily be converted for use in all of the circuits.)

The material provided in these notes was never intended as an introduction to microelectronics, but as a resource to be called on in product design work. But, perhaps for the reasons enumerated above, it has proved an effective way into microelectronics for many CDT specialists, and it is to be hoped that what follows – together with some additional reading[1] – will be of value in opening up opportunities for microelectronics in mainstream CDT work.

Each set of notes is reprinted in its original format. A discussion of the basic circuitry is followed by suggestions for applications, but cross-references to other parts of the book have been added as appropriate. It should be noted that the *practical circuit diagrams are part-pictorial and part-schematic for ease of interpretation by the beginner.*

Introductory notes on CMOS logic

An integrated circuit (IC) is an electronic device consisting of a number of components built into a single structure. Many of these, such as transistors and resistors, can be incorporated into a single IC at a fraction of the cost of the separate (discrete) equivalents required to provide particular circuit functions. All the circuits discussed here are based on the use of one particular logic IC in the CMOS family, but it would be appropriate to look at some broader principles before examining this one in detail.

Logic gates

Fig. 92 shows the conventional (American) symbol for the most simple type of logic gate: the NOT gate or inverter. The symbol actually represents a complete circuit – the details of which will not concern us here. (It is normal to adopt a black-box approach,

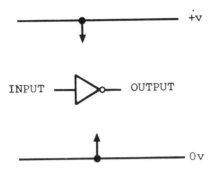

92 *Invert-gate: American symbol*

and treat the gates as complete functional entities.) The gate is illustrated, as convention dictates, between two supply rails taken, for example, from a battery (with negative shown here as 0v).

(It makes economic sense to build several logic gates into a single IC, and so it must be imagined that this is one of several identical circuits housed in the same package. The arrows shown in fig. 92 and subsequent theoretical circuits indicate power supply to the gate(s). In a typical IC having two rows of seven pins, pin 7 (0v) and pin 14 (+v) are normally used for current supply to the gates, but it should be stressed that although all the gates derive their supply current in common from these two pins, the gates otherwise operate as independent entities.)

A logic-gate circuit can have only two possible output states; it can be said to be either HIGH or LOW (corresponding to binary notation 1 and 0 respectively). This means, in effect, that the output can be either +ve or −ve with respect to the supply rails. Assuming conventional current flow (i.e. from +v to −v), when the output of the gate is HIGH, current will flow from the gate output to 0v through, for example, an LED (light emitting diode) connected between the two (fig. 93). If, when the output is HIGH, the LED is now connected across from the output to the +v rail (fig. 94) nothing happens since both ends of the LED are effectively connected to +v. If the gate output is now changed to LOW, the LED shown in fig. 94 will light up since current now flows from the +v supply rail through the LED to the gate output. But if, when the output is LOW, the LED is connected as in fig. 93, nothing happens since both ends are now effectively connected to 0v. (At this point the question often arises: Where does the gate output source current from or sink it to? The answer is that all gates on the IC are already connected to the

100

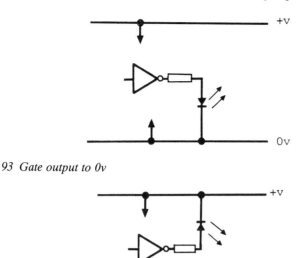

93 *Gate output to 0v*

94 *Gate output to +v*

two supply rails through the two common supply pins on the IC package, and any current involved in a sinking or sourcing transaction at the gate output(s) will pass through a gate circuit via these connections.)

What causes the gate output to go either HIGH or LOW? In the case of our invert-gate, it is simply by taking the input of the gate HIGH or LOW, and this is achieved by connecting it to either the +v supply rail (HIGH) or the 0v supply rail (LOW) – using, for example, a length of wire or a resistor. The inverter derives its name from the fact that when the input is taken HIGH, the output goes LOW, and when the input is taken LOW, the output goes HIGH; i.e. an inversion takes place.

The relationship between the input and output of a logic gate is shown in a *truth table* which normally uses binary notation as illustrated in fig. 95. In terms of the foregoing explanation, the terms HIGH and LOW could be substituted for 1 and 0 respectively.

Other types of logic gate perform in more complex ways. The truth table shown in fig. 96, for example, is that for a dual-input NOR gate, and it shows the range of options for getting the output of the gate either HIGH or LOW (i.e. three options for

101

Input	Output
0	1
1	0

95 *Truth table for invert-gate*

96 *Truth table for dual-input NOR gate*

Input		Output
0	0	1
0	1	0
1	0	0
1	1	0

getting it LOW and only one for getting it HIGH). Truth tables reveal logical relationships at a glance, and in describing these we might use the kind of language favoured by logicians – e.g. the output of the NOR gate is HIGH *when and only when* the two inputs are LOW.

Comparison of the truth tables shown in figs 95 and 96 reveals that a dual-input NOR gate can be turned into an inverter simply by connecting the two inputs together. This kind of conversion can be performed with many ICs – an especially useful facility when you have a stock of ICs not of the type required but having conversion potential.

Integrated circuit 4001

All the circuits put forward here use the CMOS IC 4001[2] which comprises four dual-input NOR gates. In all cases, these are converted to inverters by the expedient of connecting the two inputs on each gate together. The question then naturally arises: Why not use an IC offering invert-gates in the first place? The answer is to do with both performance and cost; IC 4001 is highly adaptable and remains cheap in spite of the current crisis in IC supply. (Note: all *theoretical* circuit diagrams shown use the inverter symbol rather than a NOR gate with inputs wired together.)

In common with many ICs, the 4001 is housed in a 14 pin DIL (dual-in-line) package, e.g. one having two parallel rows of seven pins. Fig. 97 shows the top view of IC 4001 with gate symbols added. This is a standard method of providing information on ICs and indicates gate type(s) as well as precise information on pin connections at a glance. Textbooks very often include IC surveys in the form of an appendix, giving information on a range of ICs in the above form. More extensive information can be obtained through specific data handbooks.[3]

102

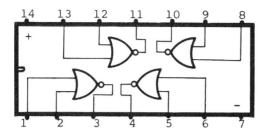

97 *Pin connections for IC 4001*

In order to convert all four gates of IC 4001 to inverters, it is clear from looking at fig. 97 that the following pairs of pins must be connected: 1 & 2; 5 & 6; 9 & 8; 12 & 13. In most of the following circuits, only one or two gates are used; this may sound wasteful, but ICs are remarkably cheap for what they contain. It is important, however, not to leave unused gates 'floating'; the inputs of such gates should be connected to the most convenient supply rail – e.g. the diagrams here all show pins 8 & 9 and pins 12 & 13 (inputs of gates not used at all) connected to the +v supply rail. Failure to 'tie' them down in this way will otherwise result in increased current consumption. It is equally important that the outputs of these gates are *isolated*.

To carry out the experiments shown in figs 93/94 (above), IC 4001 is first connected to a supply such as 9V PP3 battery via pins 7 and 14. The two input pins of the selected gate are then attached to a flying lead which can be offered to either of the battery terminals, and an LED is connected from the output pin to the battery as appropriate (remembering to use a series resistor).

CMOS ICs

There are two principal families of logic IC, each of which is constructed in a different way; these are TTL (transistor-transistor-logic) and CMOS (complementary metal-oxide semiconductor). A CMOS or TTL version of most logic ICs is available, but CMOS was chosen for exclusive use here because, unlike TTL, it can be operated over a wide voltage range. It also draws very little current, which makes it ideal for the electronic switch circuit.

Unfortunately, CMOS is prone to damage through static

103

discharge during handling. Although internal circuitry now offers some protection, it is prudent to take precautions before touching the pins. Any build up of static due to rubbing of clothing, etc. can be eliminated by touching an earth source such as a water pipe. For protracted handling of CMOS devices, hand contact can be maintained on an earthed metal (or anti-static) sheet placed on the surface of the bench or table. (ICs are also prone to heat damage, and it is usual practice to use IC sockets rather than solder pins directly.)

In spite of these warnings, it has to be said that CMOS logic ICs appear remarkably durable and difficult to damage. They can be operated over a wide range of temperatures and voltages (e.g. 3 to 15 volts), and will survive periods during which the gate outputs are short-circuited to supply rails!

Practical circuits using IC 4001

The electronic switch

The circuit of the electronic switch uses two inverters to provide an on/off latching action. When the input of gate 1 (fig. 98) is taken HIGH by connecting to +v – using a piece of wire, resistor, or even a finger to bridge across – its output goes LOW. This in turn causes the output of gate 2 to go HIGH since its input is connected to the output of gate 1. When the output of gate 2 is HIGH, current will flow to energise the LED. Because a resistor is connected between the output of gate 2 and the input of gate 1, the HIGH output of gate 2 will 'hold' the input of gate 1 HIGH even when the original connection from the input of gate 1 to +v is removed. Hence the description: latching action. Similarly, if the input of gate 1 is now taken LOW by bridging across to 0v, the output of gate 1 goes HIGH causing the output of gate 2 to go LOW; the resistor then 'holds' the input of gate 1 in this state, and the LED remains 'off'.

The practical circuit shown in fig. 99 is identical in principle to the theoretical one in fig. 98. It uses two of the four (NOR) gates available on IC 4001 which are converted as inverters by connecting together the two inputs in each case. Pins 1 & 2 are therefore connected to form the input of gate 1. The output of this gate (pin 3) is connected to pins 5 & 6 which together make up the input of gate 2 – whose output is pin 4.[4]

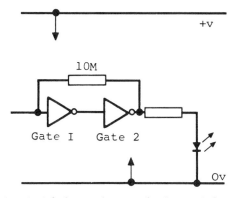

98 *The circuit in principle for an electronic latching switch*

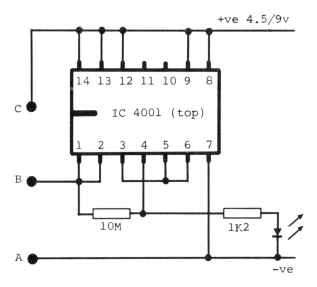

99 *A practical circuit for the electronic switch using a pair of dual-input NOR gates with inputs connected to give inverter functions*

Variations on a theme
1 If the LED (fig. 99) is connected between the output of gate 2 and +v, the switching action of the circuit is reversed.
2 The circuit can be turned into the equivalent of a spring-loaded push-button switch by taking out the 10M resistor and connecting it between the input of gate 1 and 0v. This can now

be described as a pull-down resistor since it keeps the input of gate 1 LOW (and therefore the output of gate 2 LOW). When the input of the first gate is also connected to +v using a conductor of a lower resistance than the pull-down resistor, the input is taken HIGH but only for as long as contact is maintained. As soon as it is removed, the pull-down resistor takes the input LOW again. It is probably already clear that the value of the pull-down resistor is important; for most purposes, we should want it to be as high as possible (e.g. 10M), so that even a finger bridging across from the input of gate 1 to +v will temporarily overcome its pulling-down action. By altering the value of the pull-down resistor, we can thus adjust the sensitivity of the switch. (Note: for the push-button action just described, the required effect can be achieved using just one invert-gate.)

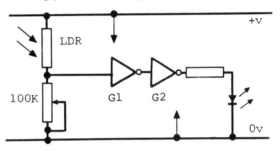

100 Light-controlled switch

3   Using the potential divider arrangement shown in fig. 100 to determine the input state of the first inverter, a light-dependent switch can be created. In the example, one resistor is an LDR (light-dependent resistor) and the other an adjustable type. The resistance of the LDR is high when not illuminated, but this drops considerably in bright light. If the variable resistor is set to just pull the gate input LOW at a certain level of illumination, when this is increased the resistance of the LDR will fall to pull the gate input up HIGH. The light sensitivity of the circuit thus depends on the adjustment of the variable resistor. (For CMOS, the changeover or transfer voltage at the gate input is half the supply voltage – approximately 4.5v if a PP3 battery is used.) It might appear at first sight that the second inverter is superfluous here since the LED could be connected between the output of the first gate and either +v or 0v. In fact, it does

the very useful job of sharpening the response of the circuit to produce a much crisper switching action. (Experiment with and without to determine the difference.)

If a latching action is required for this particular circuit, there are several possible options. For example, a latching resistor can be connected between the two gates as in the original switch circuit. In this case it should also be adjustable, and it is convenient to make it the same value as the other variable resistor.[5] Since adjustment here requires some skilful 'fine tuning' of resistors, an easier alternative method of latching might be to use a suitable SCR in conjunction with the basic circuit. (See introductory reading for information on SCRs.)

4 Substituting a thermistor (heat dependent resistor) for the LDR converts the basic switch into a heat dependent one, and similarly allowing moisture to bridge across two conductors in place of the LDR provides a moisture/water/rain detector. In all these cases, however, experimentation with component values is called for.

The electronic switch in CDT

The discussion so far may well have suggested a number of possible uses in product design work. Some examples of applications include:

(a) Touch switch control
The circuit can be used for simple on/off control either by bridging across a set of input touch points (fig. 99) directly with a finger or by taking the three input leads to a membrane panel switch (see section 1). If this is constructed using card or paper, graphite lines drawn heavily with a soft pencil may be substituted for the aluminium foil suggested earlier.

In the 'off' mode, the current consumption of the electronic switch is negligible, and the battery will almost certainly fail through 'old age' rather than quiescent current drain. Unfortunately, the current output capability of the IC is very limited, and where it is required to switch an appreciable load (e.g. a small electric motor or filament bulb), an extra stage in the form of a power transistor must be added as shown in fig. 101. It should be noted that when an inductive load such as a motor or relay electromagnet is connected, a diode should be added to avoid high voltage spikes damaging the transistor. The resistor

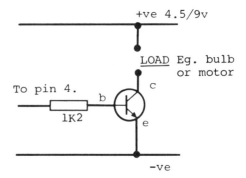

101 Transistor output stage for the electronic switch

limits the base current of the transistor to an acceptable level, and can be varied to suit the circumstances.

(b) Alarm systems
Because of its latching action, the switch circuit is an ideal starting point for all kinds of alarms. The briefest contact between +ve and the input of gate 1 will take it HIGH and cause the switch to latch in this condition – energising either an LED or, perhaps, a low-current buzzer unit. The use of a larger electromagnetic warning buzzer or bell would entail the addition of a power transistor as shown in fig. 101.

There are countless ways of momentarily taking the input of gate 1 HIGH to trigger the alarm (and taking it LOW again to turn it off). These include:

1 *Membrane panel or mat switches* These switch types are discussed in section 1, and can be constructed very easily using card, plastic sheet or paper, and foil or graphite lines for conductors.

2 *Reed switches* Reed switches are commonly put to use in alarm systems since they are closed by the proximity of a magnet and not physical contact. This means, for example, that the magnet can be buried in a door and a reed switch concealed in the frame adjacent to it. A reed switch connected between 0v and the input of gate 1 can also be used to turn the alarm off. Concealed in the alarm unit (or elsewhere), it will thus require a magnetic 'key' to be brought into close proximity.

3 *Tilt switches* A very cheap and reliable tilt switch can be

created by setting up a reed switch in close proximity to a pendulum-mounted magnet. This can be a useful starting point in the solution of a problem such as the design of an audible warning device to indicate excessive tilt on farm or building construction vehicles. (This is a very real problem, and the cause of many serious industrial accidents.) An alternative version of the switch might consist of plain mechanical contacts positioned in relation to a small pendulum (e.g. a length of weighted schim brass) which itself completes the circuit if it touches one of these points.

Where total inversion is required to switch on or off, reed switches can be secured to either end of an old pen barrel so that a magnet running inside will close them as it passes.

4   *Trembler switches*   A trembler switch can be constructed very easily by mounting a small metal mass (e.g. ball bearing) on the end of a compression spring and enclosing this in a metal tube so that the clearance between the two is quite small. The spring and tube might be mounted on an acrylic base to insulate the two.

(c)   Games and toys
The electronic switch provides an ideal opportunity for updating and extending traditional continuity-type electric games where, for example, a probe in the form of a loop has to be carefully passed along a contorted length of wire without the two touching and completing a circuit. If the electronic switch is used as the basis for this type of game, two advantages are obvious: the lightest contact of a probe on the wire will activate the switch and it will keep the alarm signal latched on. (It is doubtful in the traditional version whether gentle collisions would have registered at all.)

More interesting, perhaps, is the thought that one or both of the input leads to the switch could be heavy graphite lines drawn on paper. The contorted wire of the older game might now become a maze of lines on paper to be 'followed' by a metal stylus connected to the switch via a flexible lead. Taking this one step further, and at the same time introducing a strong graphics element, a system of graphite conductor lines on paper might be overlaid with a second sheet on which appear the graphics for something like a treasure hunt game. Pressing a sharpened probe (connected as before to one side of the switch input via a flexible lead) through the overlay at the strategic points would thus complete the input circuit through the graphite lines below. To make good contact between the graphite lines and connecting

Microelectronics for product design

lead, stranded wire should be used and fastened to the paper with staples; this provides good electrical contact as well as secure mechanical anchorage.

It is probably worth concluding with a note about ball bearing games where a ball bearing coming to rest between two or more contact points completes a simple circuit. In practice, the light weight of a small ball bearing and contact resistance usually militate against such an arrangement working consistently. If in doubt, take the contact points to an electronic switch.

The timer

The timer employs a pair of inverters and a simple capacitor/ resistor network across the supply (fig. 102). The latter is a very familiar arrangement in timing circuits, and its operation here simply involves the capacitor charging up via the resistor until the voltage at the junction of the two reaches the transfer voltage of the first gate. (The transfer voltage for CMOS is half the supply voltage – i.e. approximately 4.5v if a PP3 battery is used.)

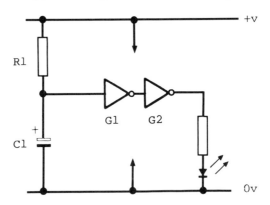

102 The circuit in principle for a simple timer

If we assume that the capacitor is fully discharged at the start of the timed interval, it has the effect of pulling down the first gate input LOW. The capacitor then charges up via R1 eventually reaching the gate transfer voltage and thus taking the input of gate 1 HIGH. The output of gate 1 therefore goes LOW and causes the output of gate 2 to go HIGH – at which point the LED, shown as an example, will be energised. The time taken

110

for the capacitor to charge up to the transfer voltage – i.e. the timed interval – depends on the values of R1 and C1, and is infinitely variable if R1 is adjustable.

The timer can be assembled using just one inverter, but adding the second sharpens up the response of the circuit to give a crisper switching action. It also means that an LED (or low-current buzzer) returned to 0v from the output of gate 2 signals at the end of the timed interval, but this is only one of several possible options. (What happens, for example, when the output is taken from a third inverter?)

In practice, the timer shown in fig. 103 is based on IC 4001 with two of its four NOR gates wired to provide inverter functions – i.e. the inputs of the first gate (pins 1 and 2) are connected together as are the inputs of the second gate used (pins 5 and 6). (The remaining gates are disabled by taking their inputs to +v.) Referring to fig. 103, it will be seen that the SPDT switch has a dual purpose; when the switch is in the 'off' position, it also short circuits the capacitor so that this is fully discharged after each timing cycle. The additional 3K3 resistor between the capacitor/resistor junction and the input of the first gate is put in to eliminate any risk of the charged capacitor attempting to

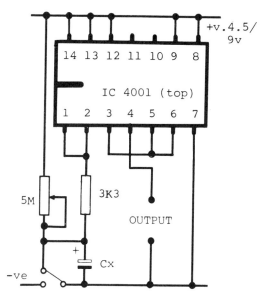

103 A practical timer circuit using two dual-input NOR gates with inputs connected to give inverter functions

power the system through the IC input protection circuitry if the supply is disconnected but the capacitor not discharged.

The component values given in fig. 103 can, of course, be altered to suit particular applications. If, for example, the timed interval is to be only a matter of seconds at the maximum setting, it makes sense to use a lower value variable resistor, which would give far better adjustment control. The two component values shown will give a maximum interval of some four minutes, but in seeking to prolong this (e.g. by using a larger capacitor), the weaknesses of the circuit begin to be exposed. If the timer is required to switch heavier loads than an LED or low-current buzzer, the output from pin 4 should be taken to an additional power transistor stage (fig. 101) using either the NPN transistor shown or a comparable PNP type – depending on what is required of the timer.

Occasionally, one type of indication is required during the timed interval, and another at its expiry. The theoretical circuit shown in fig. 104 illustrates how this can be achieved – using different coloured LEDs for illustration. A third inverter with a green LED output is added to the circuit shown in fig. 102. During the timed interval, the input of gate 2 is HIGH and its output is therefore LOW – causing the red LED to remain 'off'. At the same time, since the input of gate 3 is taken LOW by the output of gate 2, its output is HIGH and the green LED is 'on'. As soon as the transfer voltage is reached at the input of gate 1, the whole situation reverses: the output of gate 2 goes HIGH causing the red LED to come 'on', and the output of gate 3 goes LOW – extinguishing the green LED.

In practice, any additional gate on IC 4001 can be used to make this modification. For example, referring again to fig. 103, pins 8 and 9 are the inputs to a gate whose output is pin 10. Making sure that pins 8 and 9 remain connected together, break their connection to the +v supply rail and connect instead to pin 4 (the output of the second gate). The second (green) LED is then connected from pin 10 to the 0v rail.

Timer applications in CDT

The problem of timing short intervals can extend into many interesting and instructive assignments. The following list, by no means exhaustive, suggests some possibilities:

112

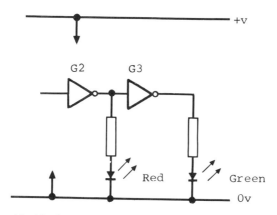

104 Timer with either/or output configuration

1 *Cooking* Design and manufacture of an adjustable timer for intervals of up to 4 minutes.
2 *Photography* Design and manufacture of an adjustable timer for intervals of up to 20 seconds.
3 *Athletics* Design and manufacture of a pre-set 'target' timer which calls for maximum performance (number of press-ups?) in a limited time.

All of these problems entail a wide range of considerations including: a precise formulation of the design brief, choice of circuit details and component values, design of the physical package and choice of signalling and timed interval adjustment. **Note: For a system of timing over longer periods, see the counter/timer notes below.**

Multivibrators

The basic multivibrator circuit (fig. 105) uses a pair of inverters to provide, in effect, an on/off switching action the frequency of which is determined by the values of R1 and C1. Represented graphically, the output of the circuit is a square wave – or, at least an approximation to one (see fig. 112). The most effective way of describing the operation of this circuit is sequentially as follows:

(a) We will begin by assuming that the instant the circuit is switched on, the output of gate 1 is HIGH – and therefore

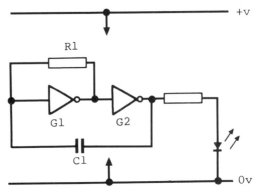

105 The circuit in principle for a multivibrator

the output of gate 2 is LOW. (In practice this is a matter of chance.) In this state, C1 will charge up via R1 until the voltage at the junction of C1 and R1 has risen to the transfer voltage of the input to gate 1.

(b) When this transfer voltage is reached, the input of gate 1 goes HIGH and its output goes LOW – causing the output of gate 2 to go HIGH. When this changeover occurs, C1 is discharged.

(c) C1 now charges up via R1 until the voltage at the input of gate 1 falls to the transfer voltage – at which point its input is taken LOW and its output goes HIGH, taking the output of gate 2 LOW. At this instant of changeover, C1 discharges again and the whole cycle begins once more from (a) above.

The operating frequency of this circuit can be varied through a very wide range; increasing the resistance of R1 and/or the value of C1 reduces it – and vice versa.

In practice, two of the gates on IC 4001 are used as inverters, and the remaining gate inputs – pins 8 & 9 and pins 12 & 13 – are taken to +v to avoid wasted current (fig. 106). The component values indicated will give a relatively low frequency appropriate for experimenting with the LED shown connected to the multivibrator output. The highest frequency at which individual 'flashes' are normally discernible, is governed by persistence of vision, and is actually quite low. Increasing the frequency beyond this limit will cause the LED to appear 'on' all the time, and this is often mistaken for a fault in the circuit. Similarly, the frequency can be slowed down to a point where the lengthy on/off dwell times give the impression that the circuit is not

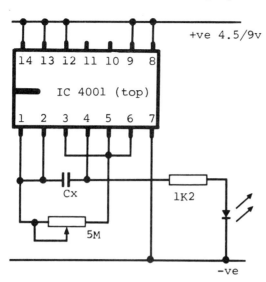

106 *A practical multivibrator circuit using a pair of dual-input NOR gates with inputs connected to give inverter functions. Capacitor Cx could be 47μF or 100μF for the LED output*

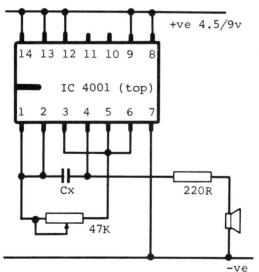

107 *Multivibrator: a practical circuit with loudspeaker output. The normal audio frequency range can be covered by inserting the following range of capacitors for Cx – .0047μF, .04μF, .47μF. The loudspeaker should be a low impedance miniature type*

115

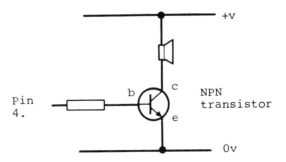

108 Transistor output stage for the multivibrator

functioning at all. (It may be of interest to know that the ability to perceive or resolve individual 'flashes' at higher frequencies appears to depend partly on our state of fatigue; the greater this happens to be, the poorer our capacity for resolution.)

A small loudspeaker can be substituted for the LED to give an audible output (fig. 107). The sound level will be modest because of the low current output capability of the IC; however, providing the loudspeaker with a baffle (even just a piece of card with a hole cut into it) or, better still, putting it into an enclosure, has a remarkable effect on the volume. If a louder sound is required, an additional output stage in the form of one transistor can be added as shown in fig. 108. This can be any one of a large number of general purpose NPN transistors – e.g. BFY51 or the higher power TIP31A. The resistor shown in the diagram limits the transistor base current to an acceptable level, and a variable resistor (log. type) could be added at this point to provide a volume control.

Multivibrator applications in CDT

The multivibrator circuit can be used and adapted for a surprisingly wide range of applications, and these will be dealt with under the following headings: *Optical and audio applications*; *timing and counting*; *motor control*.

1 *Optical and audio applications*
(a) *Signalling or warning indicators*   The basic circuit with LED output can easily be incorporated into an alarm system to provide visual warning. A more effective method of drawing attention is

to arrange for two alternately flashing LEDs – which only requires an additional LED between the output of gate 1 (pin 3) and the 0v supply rail.

(b) *Stroboscope* The circuit produces a very regular (if somewhat imperfect) square wave output, and if calibrated to sufficient accuracy against a reliable reference frequency, it could itself be used as a reference device – perhaps to check the speed of cheaper record turntables, i.e. in conjunction with a segmented disk resting on the turntable. It can be similarly used to check the speed of any revolving component, but in all these applications it would help to substitute a high-intensity LED (possibly switched by a transistor output stage).

(c) *Freeze-frame device* This is a possible application of (b) above which might lead to the design of optical toys. If a flashing (high-intensity) LED is synchronised with the speed and offered to the surface of a rotating disk, a series of sequentially-altered images drawn around its periphery will be momentarily frozen each in turn – giving the appearance of movement to whatever image theme is chosen. (It is an advantage here to be able to alter the mark/space ratio of the output – see *motor controller* notes below.)

(d) *Audio signalling* The multivibrator can be used for simple signalling or warning, and since its frequency is variable, it has considerable potential here. For example, most people who suffer from loss of hearing typically suffer a loss of response to *particular* frequencies – especially the higher range. An interesting and worthwhile project might therefore be the design and manufacture of a signalling device (doorbell or telephone sounder?) for use by a partially deaf person. This would involve a number of interesting problems including determination of the optimum frequency for a particular person.

(e) *Signal generator* As a simple signal generator, the multivibrator can be useful in several areas of design work. It might be used, for example, to give an acoustic *indication* of resonance problems in enclosures such as loudspeaker cabinets, or it could be used in noise level experiments in association with a noise level meter. For those with a special interest in physics, there is always a need for apparatus in school to demonstrate different acoustic phenomena.

(f) *Musical toy* Because of its square wave output, the sound of the multivibrator is quite strident, but it can be turned into something like a one octave organ simply by switching in – via a suitable keyboard – a number of pre-set resistors each adjusted to give a particular frequency. As a possible design and make project, a musical organ for younger children offers a lot of scope if the keyboard is designed and constructed as a membrane panel (see *section 1*).

2 *Timing and counting*

Since its output is quite regular, the multivibrator can be put to use as a *clock* base for counting and display-driving circuitry. We are fortunate in having a CMOS IC (4026) which provides both counting and display-driver functions in one convenient package, and so it is relatively easy to produce a multi-digit display timer or an ordinary LED display clock.

IC 4026 (which comes in a 16-pin dil package) will count the number of square wave output pulses from the multivibrator and display this count as a digit on a seven segment LED display. If more than one digit is required – e.g. for counting up to 99 or 999 – one additional 4026 is added to drive each display. Pin 5 on IC 4026 provides a divided-by-10 output for connection to the next IC, and when the first digit in a multi-digit display goes from '9' back again to '0', a single pulse is sent to the next IC which consequently counts '1' and shows this on its display.

Before looking in detail at how to connect IC 4026, it might be useful first to consider how the digits are created. A seven segment LED display is an arrangement of LEDs juxtapositioned in a single unit so that they can represent any number between '0' and '9' when they are lit up in the appropriate combination (fig. 109). There are seven separate LEDs which, for convenience of packaging and connecting, have all their anode connections (or, conversely, cathode connections) taken to a common pin on the package. In order to get a particular segment of the display to light up, the common pin must be connected to the correct supply rail (+v for common anode and 0v for common cathode) and the other side of the supply to the appropriate pin via series resistor. Simply inspecting a typical seven-segment display gives no help in identifying which pin belongs to which segment, but by common agreement each segment is given a letter (fig. 109), and data sheets for particular display types (and some supply catalogues) will clearly show which lettered segment belongs to which pin.[6]

If an unidentified seven-segment display is encountered, the common pin (anode or cathode) together with the others can

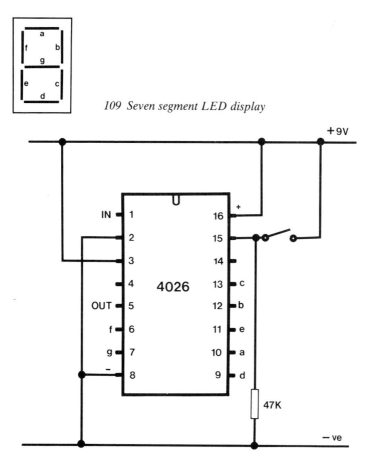

109 Seven segment LED display

110 Wiring diagram for IC 4026

quickly be sorted out by trial and error using a battery and series resistor. (Some displays are based on CMOS technology, and are thus prone to damage by static discharge.)

IC 4026 is designed to drive a common cathode seven-segment display, but unlike many other devices it does not call for the use of series resistors when connecting to the LED segments. The pins shown in fig. 110 marked a-g are therefore connected directly to the corresponding pins on the seven-segment display. Other connections should be made as indicated on the diagram. The multivibrator output (pin 4) is taken directly to the clock input pin (1). Pin 15 is the counter reset input, and is here pulled LOW by R1, but can momentarily be taken HIGH by closing the switch. This action sets or resets the display to zero. (Alternatively, a small capacitor can be substituted for the switch to give a short positive pulse when the circuit is switched on.) Pin 3 is shown taken HIGH by connecting it directly to the +v supply rail, but if this is taken LOW – say, with an appropriate switch – the circuit continues to function normally but the display is not driven – thus conserving current over a period of time. Finally, pin 5 is the divide-by-10 output which can be directly connected to the input pin of the next 4026.

IC 4026 offers several other facilities not indicated on the diagram, and the interested reader is referred to any appropriate data book for details of functions and pin connections.

The combination of multivibrator *clock* and seven-segment display via IC 4026 opens up many opportunities for product design work in CDT where a timing and display function is needed. The clock can be adjusted over a wide frequency range, and the entire system can be controlled in a variety of ways which invite experimentation and innovation.

It is perhaps worth noting, in conclusion, that IC 4026 can also be used in the design and construction of distance measuring and recording devices. Since it counts each LOW to HIGH transition of the square wave from the clock, a conventional switch that repeatedly takes the input to pin 1 LOW and then HIGH can be substituted. If such a switch is attached to a form of trip wheel (or made part of the wheel itself), the digital readout will represent distance covered as the wheel revolves. In practice, the measuring device might be for small-scale work (e.g. opisometer for map measurements) or for large-scale measuring (e.g. a pedometer). Somewhere between the two, an interesting problem is the design and constructing of a DIY measuring tool – e.g. a device for measuring lengths of wallpaper to the nearest centimetre. (The problem of mechanical switch bounce causing erroneous readings can be eliminated by including a simple de-bouncing circuit constructed from another logic IC.)[7]

3 *Motor control*

The speed of small DC motors can be controlled in a variety of ways – the most obvious of which is to connect a wire-wound variable resistor in series. This is the system employed in cheap model train controllers, but the big drawback is that motor torque drops disproportionately at low running speeds.

One method of achieving better torque characteristics at low

120

speeds is to rapidly switch the supply on and off and simply vary the ratio of the time 'on' to the time 'off' – i.e. by controlled pulsing of the supply current. The circuit shown in fig. 111 is designed to do this, and is based on a modification of the basic multivibrator. The original circuit produced, in effect, a high speed on/off switching action with the 'on' period roughly equal to the 'off' time. Represented graphically, it appears as the square wave shown in fig. 112 and is said to have an equal mark/space ratio – i.e. equal times 'on' and 'off'.

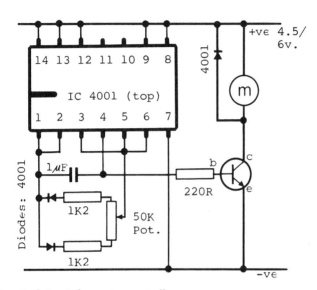

111 Practical circuit for motor controller

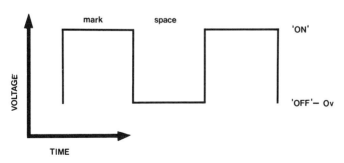

112 Ideal square wave

113 Square waves with altered mark/space ratios

In controlling motor speed, we need to adjust the mark/space ratio so that the 'mark' time is relatively short to give slow running and relatively long for fast speed (fig. 113). This is achieved in the modified circuit by using a pair of *steering* diodes which control the charging of the capacitor via the variable resistor – which is now effectively two resistors each coming into operation alternately. A further consideration of the action of the multivibrator (see above) should make this clear.

It is not possible to run a motor directly from the IC output since there is very little current available. An additional stage is therefore added once again in the form of a power transistor which could be any general purpose medium to high power NPN type. (As the motor is an inductive load, a protection diode is added.) The drawback of this circuit – or, rather, the principle itself – is the effect of pulsing on the smooth running of a motor, especially at low speeds. But used to control inexpensive motors for non-precision work, this is certainly the lesser of two evils.

The controller circuit can be used in many applications requiring infinitely variable speed control of a small motor. An interesting and worthwhile product design exercise is the design and construction of a controller for an electric slot-car or train set (assuming a safe 12v supply from an approved transformer unit). In addition to speed control, such a device would need to incorporate a positive on/off switch as well as a reversing control in the case of a train.

Notes

1 For some introductory material see: A. Hedley, *Electronics: Data Design & Construction*, Harrap, 1983. A. Hedley, *Electronics: An Introduction*, Harrap, 1983. Schools Council, *Basic Electronics (Books 1-5)*, Hodder & Stoughton, 1978. Schools Council, *Modular Course in Technology: Electronics*, Oliver & Boyd, 1981, etc. J. Watson, *Mastering Electronics*, Macmillan, 1983. R.A. Sparkes, *Electronics for*

Schools, Hutchinson, 1972, etc. M. Sladdin, *Elementary Electronics*, Hodder & Stoughton, 1981. G. Bevis and M. Trotter (eds), *Microelectronics: Practical Approaches for Schools and Colleges*, MEP/BP. For more advanced reference material, see, for example: P. Horowitz and W. Hill, *The Art of Electronics*, Cambridge University Press, 1980.

2 Note: 4001 is the generic type number. Available ICs will probably be in the 'B' series and marked 4001B. They may also display code letters denoting maker's name and date of manufacture.

An excellent supplier of most components discussed is: JPR Electronics, Unit M, Kingsway Industrial Estate, Bedfordshire, LU1 1LP. But please note: this is a wholesale supplier and the minimum order is £20 as at 1985.

3 For an example of a short CMOS survey see: *Electronic Games*, Elektor Publishers Ltd, 1983. There are many comparable examples of CMOS surveys included in texts as an appendix. For more comprehensive surveys and data see: D. Lancaster, *CMOS Cookbook*, Howard Sams & Co. Inc., 1977. C.L. Hallmark, *The Master IC Cookbook*, TAB Books Inc., 1980. (Although both American books, these publications are widely available through bookshops in this country.) A definitive reference guide is: *D.A.T.A. Book: Electronic Information Series. Digital Integrated Circuits*, vol. 28, Book 33, Nov. 1983.

4 Since there are four gates on IC 4001 – two of which remain unused for the basic electronic switch – each IC can, of course, be used to provide two switching functions.

5 To get the circuit working, first of all set the latching resistor to its highest value and adjust the potential divider to obtain the required light sensitivity. The latching resistor is then adjusted until it just provides the latching action needed – which it will do providing its resistance is less than that of the variable resistor in the potential divider.

6 See, for example, the RS Catalogue: RS Components Ltd, PO Box 99, Corby, Northants, NN17 9RS.

7 See, for example, Horowitz and Hill (note 1 above), p. 387.

Index